D1132622

A GENEROUS NATURE

Advice from a Raindrop

You think you're too small
to make a difference? Tell me
about it. You think you're
helpless, at the mercy of forces
beyond your control? Been there.

Think you're doomed to disappear,
just one small voice among millions?
That's no weakness, trust me. That's
your wild card, your trick, your
implement. They won't see you coming

until you're there, in their faces, shining,
festive, expendable, eternal. Sure you're
small, just one small part of a storm that
changes everything. That's how you win,
my friend, again and again and again.

KIM STAFFORD, OREGON POET LAUREATE

A Generous Nature

LIVES TRANSFORMED BY OREGON

Marcy Cottrell Houle

Oregon State University Press   Corvallis

Library of Congress Cataloging-in-Publication Data

Names: Houle, Marcy Cottrell, 1953- author.
Title: A generous nature : lives transformed by Oregon / Marcy Cottrell
    Houle.
Description: Corvallis : Oregon State University Press, 2019. | Includes
    bibliographical references and index.
Identifiers: LCCN 2019035594 | ISBN 9780870719790 (paperback)
Subjects: LCSH: Wildlife conservation—Oregon. |
    Conservationists—Oregon—Anecdotes.
Classification: LCC QL84.22.O7 H68 2019 | DDC 333.95/41609795—dc23
LC record available at https://lccn.loc.gov/2019035594

♾ This paper meets the requirements of ANSI/NISO Z39.48-1992
(Permanence of Paper).

Oregon State University
OSU Press
Oregon State University Press
121 The Valley Library
Corvallis OR 97331-4501
541-737-3166 • fax 541-737-3170
www.osupress.oregonstate.edu

To Lucy, Lilly, and our future grandchildren,
all seventh generation Oregonians.

I pass the mantle to you.

Everything that has gotten done in the past
is because somebody cared. They cared
passionately, often devoting much of their life
to it. New issues are emerging all the
time. Attacks are being made to roll back
achievements of the past. It will be never
ending. It takes new generations, succeeding the
others, to pick up the mantle and to carry the
cause forward. It requires people who believe
in themselves and understand that they can
make a difference. Never underestimate the
influence that an individual can have.

―――――――――

J. Michael McCloskey

Contents

Introduction

This book is a love song to Oregon. It is not mine alone, but that of many who feel as I do that we are privileged to live in one of the earth's most exceptional places.

Anyone who has traveled through the dramatic basalt cliffs of the Columbia River Gorge, encountered Oregon's 362 miles of clean, windswept public beaches, observed the colorful wildflowers in the Eagle Cap Wilderness, and stood humbled by the old growth trees of Portland's Forest Park cannot help being awed by Oregon's beauty and majesty. When we enjoy the sparkling waters of the free-flowing Metolius and Rogue Rivers, and witness miles of rich agricultural lands not covered over by urban sprawl, we begin to understand that Oregon is different from any state in the country.

Most of us can appreciate these things, but something else is often overlooked. Preserving these iconic features didn't just happen. For each, it took the power of an individual to save these places for the rest of us. Without the vision and courage of these persons—the majority of whom have names that are unknown and unsung—the Oregon we enjoy today would not exist. Their dedication to Oregon has given all persons gifts beyond measure. Even more, many of the achievements originating in Oregon have informed and created environmental movements nationwide.

From Oregon's unique Bottle Bill that dramatically reduced litter to its Beach Bill that made all Oregon beaches public with "free and uninterrupted use," from the creation of the largest scenic area in the United States to the passage of innovative land-use laws designed to preserve important agricultural and forest lands and reduce urban sprawl, Oregonians have demonstrated leadership and bipartisan progressiveness that continue to influence our state and nation today.

Of course, this could change. Throughout the country and world, environmental laws are being relaxed and protections for our natural and cultural sites are being loosened. At the same time, it is easy to take for granted that the things we cherish will stay the same and always persist.

They won't. Not unless someone steps up to care. That's why the stories of people who devoted their lives to protecting a place they loved matter. Newcomers and longtime residents alike need to know them. They tell us: This is what we have. This is how it came about. These are people who made it happen.

The problem is, most of these stories are disappearing.

I learned this while serving for twenty-five years on the board of the Oregon Parks Foundation, a volunteer organization for the promotion of parks and natural areas statewide. I heard stories of people who had done so much to preserve the magnificence of Oregon. I came to realize, however, most weren't written down anywhere and were vanishing as these individuals were passing away.

There was more. I also started to see how few Oregonians, myself included, really grasped why Oregon looks the way it does. A majority do not know much about the signature laws, planning, places, and policies that have been enacted by resolute individuals who endeavored—sometimes for a lifetime—to protect Oregon's natural resources. Like many Oregonians, I heard references to many of them but could scarcely define Senate Bill 100, the Urban Growth Boundary, or the Wild and Scenic River designation. I could not really explain the significance of the Diack Act, the Coalition of Oregon Land Trusts, or the High Desert Partnership. Yet all of these helped create the Oregon we know and all are pivotal to its future.

I saw if we were to keep Oregon a place of beauty and wonder, we needed to know these things, and the stories of the people behind them. They were the tales that made these issues and places come alive! We were running out of time. And so began what turned into a ten-year journey for me to record some of them while I still could.

With tape recorder in hand, I traveled around Oregon, searching for people who had done significant things for their communities. I sat down with them in their homes or at their work or in the field to capture their words and their voices as they told me about their lives. They spoke about what Oregon meant to them and why they dedicated themselves to trying to protect it for future generations.

I learned some lessons I hadn't anticipated. People who had taken leadership roles came from all parts of Oregon, many kinds of backgrounds, and an assortment of jobs. They were teachers and ranchers and attorneys. They were fishermen and farmers and doctors. They were housewives and journalists and government officials. Some had designed extraordinary environmental education programs while others spent their lives defending and upholding land-use laws. Some persons formed land trusts, some strove to protect Oregon's agricultural heritage, while others fought to keep its rivers clean, its forests healthy, and to save endangered salmon.

Their leadership did not come out of nowhere. There is a long history in Oregon of persons working to conserve natural beauty and resources, beginning with Oregon's native peoples who lived on the land for generations and were its original stewards. For thousands of years, Native Americans showed respect for the land; they depended on it.

While everyone I interviewed revealed different approaches, there was a common thread throughout: *they loved Oregon.* Their strivings came from an inner drive for a greater good, a better Oregon, and without desire for personal gain or recognition. This last fact, perhaps, is what impacted me most of all. In recording, researching, and writing these conversations, I found myself deeply moved by something I had not expected. Even more than their accomplishments, it was their *values* that inspired me. Through all their efforts they evinced a high aim of protecting something they cared for, often devoting years to protecting something they cared for, even when the chance for success was not assured or was decidedly slim. Their motivations were not for themselves, or for something they felt to which they were entitled, rather they were doing something for others and for future generations.

These guiding values, I knew, needed to be shared too.

A Generous Nature is a collection of twenty accounts taken from my recorded conversations with people who changed the face of Oregon. While many more deserve inclusion in this book, and countless other individuals need to have their stories documented (and I hope they will), the stories I selected stand out. They are potent illustrations of participatory democracy in action. They showcase landmark actions and innovative thinking that have made a visible difference in Oregonian's lives. Their accomplishments continue to be influential today, and many have spread throughout the country. Others are acting as springboards to new programs.

In a time of global climate change, worldwide loss of biodiversity, native habitat degradation, and growing population pressures, such advocacy for resource protection is more important than ever. At the same time that federal laws cannot be relied upon to protect our environment, states such as Oregon play critical roles. What's more, they act as models others can turn to.

The quality of our lives intricately depends on how we care for the earth. That is why these stories and the inner compass that guided these Oregonians to outstanding feats of service are important to document, to remember, and to widely share. These are the stories that reveal to us the places, ideals, and laws that have made Oregon, *Oregon*.

They come, however, with a challenge. If Oregon is to retain the incomparable character these leaders helped create, it will depend on us.

It is my truest hope in writing *A Generous Nature* that the wonderful individuals I was honored to profile in this book may stir others' hearts to action.

What's in our Oregon DNA? At its core
is the knowledge that, where you are now,
is a gift to you. It is the belief that this is
one shared space for diverse people to
find commonality. It is a deep hunger to
give back what you've been given ... and to
become part of something more than just
yourself. It is welcoming every Oregonian
with: "I don't care what your background
is, what your political, your social, your
moral stripe is; I don't give a damn." Oregon
is a place where there's the big table that
everybody can sit around and look at one
another ... converse, not yell, discuss, not
complain, not ridicule, and come away from
that table saying, "This land is good.
There is goodness that defines this state.
Let's embrace that goodness."

Jack McGowan

I

Mapping the Oregon DNA

JACK McGOWAN

Rain is pelting, as it often does, on the Oregon Coast in spring. It is not deterring these numerous groups of volunteers, though. Up and down the 362 miles of public beaches of Oregon, thousands of people are smiling in the unremitting showers. They pick up broken pieces of Styrofoam, plastic bags, and food wrappers. They remove car tires and cigarette butts. By the day's end, they will have collected over eight tons of trash, leaving the beaches, owned by all Oregonians, clean and renewed, thanks to an extraordinary program called Stop Oregon's Litter and Vandalism (SOLVE), advanced by a man who claimed Oregon as his adopted home.

ᴖ

Explorers come in many forms. Some follow a craving for gold and silver, wherever that may lead them. Others hunger for new travels, growing perpetually bored of where they're rooted. Still more search paths, looking for places to leave an imprint of their own title for all to see.

Jack McGowan is none of those. His explorations were not in pursuit of fortune or fame. Rather, they were motivated by an unease that nagged him—a sense that, as a young man, his life should have more meaning, more purpose than what he was living in lower Manhattan.

An only child, he grew up in a hard-working, middle-class family without many material goods. He never had his own bedroom, but slept on a convertible couch in the living room because his folks could not afford a two-bedroom apartment. Yet, as Jack is quick to say, he had wonderful parents who loved him dearly and gave him strong ethics and a moral compass that has served him throughout life.

Raised in the middle of New York City in the 1940s and 1950s, Jack's playgrounds were fire escapes, apartment building roofs, and the streets of the city. There was no affinity, no relationship at all with the natural world. He remembers once, as a child, his parents took him to Central Park; it was the first time he had felt grass under his feet.

Even at a young age, Jack recalls feeling there had to be someplace beyond the urban ramparts and concrete environment, but the notion stayed safely vague as he rose through the ranks as a Wall Street broker. His first job was working on the floor of the New York Stock Exchange. Becoming successful, he moved from his parent's apartment to live in Greenwich Village, two train stops away from Wall Street. Within his grasp was a life of wealth, opportunity, and success. To an outside observer, and especially to his parents whom he deeply loved, Jack was on the trajectory to success.

The problem was, while basking in the pride of his parents and the promise of a successful Wall Street career, Jack could not rid from his mind a needling refrain: "There's something basically wrong here." To the despair of his parents and the chagrin of coworkers who told him he was throwing his life away, Jack decided to quit his job. Never having been west of New Jersey, he thought he'd journey to the Pacific Coast—to California.

But then something happened. On the crowded corner of Sixty-First and Sixth Avenue, an unexpected twist would change Jack's plans and forever alter the course of his life. Squeezed among the throngs waiting for the street lights to change, a quiet, unassuming man stood directly in front of Jack:

Paul Simon.

The acclaimed musician, who had recently released his smash hit "Bridge Over Troubled Water" with his friend Art Garfunkel, stood unmoving in the crush of pedestrians, awaiting his turn to cross. A huge fan of Simon's, Jack could not stop himself from saying, "I just love your music."

Paul Simon turned to Jack and, in an approachable and self-effacing manner, paused to speak to him. While the pedestrian fleet dispersed around them, not recognizing the singer, for the next twenty-five minutes Jack and Paul lingered at the street corner in amicable conversation.

"We talked about everything, from Martin Luther King and Bobby Kennedy to the Vietnam War to the March on Selma, Alabama. We talked about all the conflict—the social and ethical maelstroms that were erupting in the United States," says Jack. "And as I was explaining my plans about leaving New York, Paul said, 'Where do you want to go?'"

Jack answered that he'd never been west, that he was going to California.

"Okay. But before you go to California, may I make a suggestion?" replied Paul. "Try the Pacific Northwest. Seattle's a big city. But Portland is a beautiful, smaller town."

Jack never met Paul Simon again. But the chance street corner encounter transformed everything. At that moment, Jack chose Oregon, and from that decision he at last found the meaningful existence he hungered for.

One month later, in August 1970, Jack pulled into Portland. He didn't know a soul, found a furnished apartment to rent, put his bags down, and said, "Now what?" Three thousand miles away from home, he realized what he'd just done.

"I had no job. I had no car—because New Yorkers don't have cars. I had a modest savings left. I was beginning life all over again. And I quickly realized that all of the experiences I'd absorbed from being a New Yorker, with sometimes a hard edge, didn't play well in Oregon."

With that awareness, though, came excitement. He recognized that he had been given an opportunity to redefine himself at a young age. New experiences could help frame a different existence. Soon after arriving in Portland, he discovered a park near his residence—a forest like he had never seen before. The wonder that he felt hearing a brook, seeing an old growth Douglas fir in Portland's Forest Park, changed everything.

"I stopped and listened. I had never had that stillness before in my life," says Jack.

The silence and the immensity of the forest surrounding Jack had a profound effect. He had just stepped into the nation's largest urban wilderness park, home to more species of birds than he ever knew existed. The lush ferns, the towering Douglas fir and western red cedar, the miles of woodland trails within the five-thousand-acre city park were staggering. Forest Park, a nature preserve and crown jewel of Oregon's city parks, had claimed him.

"From my early experience in Forest Park, I started to understand, not only about my adopted land, but that Oregon in essence had adopted me. I began realizing that the gift of becoming an Oregonian also meant a huge responsibility as I became part of this good land."

Jack needed to find work. The savings he had brought from New York was starting to run out. One of his first jobs in Oregon was as a salesman at British Motorcar. The best part of that was that he was able to purchase a Land Rover—a vehicle, says Jack, that no one really wanted in 1970. But the car allowed him to do what he really aspired to: experience Oregon. On his days off Jack began to explore the state. He visited the Pacific Coast—a landscape of green, picturesque headlands and miles of unpopulated sandy beaches. He toured the basaltic cliffs and pounding waterfalls in the Columbia River Gorge. He spent time in the snow-capped mountains of Central Oregon and probed the high desert, a land of sage and juniper. The more he saw, the more he realized the varied Oregon environment was instilling a new ethos into his blood.

In 1974, four years after moving from New York City, Jack learned that KINK-FM was looking for a promotion director. It was a station he loved for its eclectic and progressive music, a combination of alternative, rock, acoustic, folk, pop, instrumental, and blues. He applied and was told there would be numerous people trying for the position. Jack took a daring approach.

"I'll tell you how badly I want this job," he appealed to the interviewer. "For a dollar, I will sign a ninety-day contract with you. After the ninety days are over, if I haven't proved that I'm the candidate you want, you keep dollar, the contract is through, and I walk away. But if at the end of ninety days, I have proved myself, then you hire me full-time and give me back that dollar."

It was a crazy proposition and Jack knew it. But he had enough money saved that he could last for ninety days and still pay the rent.

"That is the most outlandish offer I have ever had in my career," Jack remembers the manager saying. "If you're that creative in a proposition like that, then you must be one hell of a promoter. We'll get back to you."

Jack got the job.

After four years as promotion and public service affairs director, he left KINK-FM for a new opportunity as marketing director at the Washington Park Zoo, now known as the Oregon Zoo. In that role he originated several innovative programs, two that became rapidly successful: the Mt. Hood Festival of Jazz and the highly popular Oregon Zoo Concerts. The Oregon Zoo Concerts, still active today, were the first ever held in a zoological institution. The idea has now spread throughout the world.

Jack's creative initiatives were not going by unnoticed, especially by a newly elected mayor of Portland, Bud Clark, who was in desperate need of a press secretary. The mayor, who previously worked as an independent barkeeper, suddenly had the eyes of the world on him. Everyone wanted to know more about Clark, and his David and Goliath story of a local bartender successfully taking on an entrenched political system. He had to find someone who could handle all the national and international media attention deluging him. Clark had heard of Jack McGowan, and sought him out.

After listening to the mayor's appeal, Jack accepted the offer to help him with the publicity. The mayor's office knew that keeping the flamboyant mayor on track was too much for just one person—they also needed to find an executive assistant. Another employee suggested a woman he felt could actually keep up with the new mayor, and corral him, when necessary. Her name was Jan Van Domelen.

Jack remembers the first time he saw her. He was thirty-six and had never been married. When she walked into the room, he was struck by Jan's beauty. But that wasn't all. "I took one look at her and said, "I can't believe it."

He didn't know that when Jan saw him for the first time, she also thought, "I can't believe it."

Together, Jack and Jan experienced the political furnace of the ups and downs of the Clark administration. In the heat of trying to help the mayor, he and Jan forged an incredible friendship and respect for each another.

"That respect manifested into our dating, which manifested into a serious romance, which manifested into our getting married, which manifested into the beautiful son we have," says Jack, smiling.

Jack stayed with the mayor for his first term. He then received another intriguing offer, this time from the Portland NBC affiliate KGW-TV. Jack accepted the new position and became the co-host of a nightly televised show. He enjoyed the work and was very happy in his personal life. Even so, he confesses, he found himself searching for deeper meaning in his work.

"I wanted to give something back to Oregon," he says thoughtfully. "In addition to raising our wonderful son, I wished to dedicate myself to some Oregon calling."

It was during this period that Jack heard about a statewide search for an executive director for an organization he had never heard of before. It was to head up something called SOLV, which stood for Stop Oregon's Litter and Vandalism.

"It caught my attention and I started looking into what SOLV was about. There wasn't too much I could find on it, but I learned it had been founded by Tom McCall, who was Oregon's governor from 1967 to 1975. McCall left some wonderful legacies," he says. "Most people can identify three that he gave us . . . or if they can't, you can be sure they are benefiting from them. Tom McCall enacted the Beach Bill, the Bottle Bill, and Senate Bill 100. All were revolutionary and put Oregon on the world stage."

Jack admits he was highly inspired by McCall's three pillars. Each was first of its kind for any state. Each protected the natural resource and attributes in visionary new ways and enhanced quality of life. Each had a ripple effect that extended far beyond Oregon, changing the way people thought about caring for the landscape surrounding them.

Signed by Governor McCall in 1967, the landmark Beach Bill established and guaranteed public ownership of all 362 miles of the Oregon coastline. It granted persons free and unrestricted access to the entire Oregon Coast, from the water to sixteen vertical feet above the low-tide mark, from dry sands to the vegetation line. Of all US coastal states, Oregon is regarded as having the best legal protection for the public's use and access to shorelines. The Beach Bill has been called the country's model program, and one of the most far reaching measures of its kind enacted by any legislative body in the nation.

The Bottle Bill, also enacted by Governor Tom McCall, was another first in the United States. In 1971, the Oregon governor signed this historic piece of legislation requiring that cans, bottles, and other containers of carbonated soft beverages and beer carry container deposits and be returnable for a refund value. With return rates now averaging 90 percent, this law reduced roadside litter from cans and bottles in Oregon from 40 percent to a mere 6 percent. Many other states have subsequently adopted similar deposit fees.

The third groundbreaking legislation occurred in 1973 with the adoption of Senate Bill 100. Once more the first of its kind in the nation, this extraordinary land-use law—designed to protect farm and forestlands—created formal structure for statewide planning. Senate Bill 100 required that every city and county in Oregon prepare a comprehensive plan for development in accordance with nineteen prescribed state goals. These goals included priority planning for agricultural lands, forestlands, open spaces and natural resources, public facilities, energy conservation, coastal shore lands, ocean and estuarine resources, and more. It also required citizen involvement.

All three bills were incredible milestones and made Oregon a trailblazer in the nation in protecting the environment. It elevated the far-western state to an emblem for progressive ideals.

"But unknown to me and to many of us who applaud McCall's leadership, there was a fourth pillar that Governor Tom McCall also left us," says Jack.

As he was to discover while researching the position, McCall had founded SOLV in 1969. SOLV's official goal was to reduce and clean up litter, graffiti, and defacement throughout Oregon, through the staging of statewide community cleanup campaigns. It, too, was the first program of its kind in the nation, but unlike McCall's other three achievements, SOLV did not take off. Once McCall left office in 1974 and passed away in 1983, the organization lost its major champion. And by 1990, when Jack saw the job announcement, it had nearly faded into oblivion.

"I was told SOLV had no staff, no office, no phone. There was a loaned post office box, about one hundred sheets of letterhead, and $12,000 in the bank."

It was, however, the same time that Jack found himself seeking something more purposeful to do with his life—a lingering feeling that had never quite abated since the beginnings of his work life. After deep consideration and talking it over with Jan, he decided to take the job, on one condition—that his wife be hired as co-director.

The board swiftly hired them both. Soon Jack and Jan went to work moving the furniture out of their family room to make space for an office. They moved their car out of the garage. Their first year's salary, in 1990, was $10,000 with no benefits. They had a baby and a mortgage, but no fax machine until Jack spent his own salary on to purchase one.

For five years, the McGowans operated SOLV out of their home. "The job was 24–7, 365 days a year," says Jack, "and we faced some challenging economic times. The phone would ring at 10:00 on a Sunday night. One of us would race into the little office with the very old office furniture to answer it. The caller would be amazed that someone would actually answer. In that call would be an appeal from somewhere in Oregon, asking for assistance." That request often came in the form of asking for help with beach cleanups, cleaning up illegal dumpsites on national and state public lands, restoring degraded wetlands, and cleaning up local neighborhoods. Jack and Jan set them up.

Before long, the annual beach cleanup was SOLV's most well-known achievement. After several successful events, SOLV grew large enough to hire one employee—who worked from her home because Jack didn't have space in their tiny office for two desks. They had weekly Monday morning meetings and, fielding more requests, soon hired a second, half-time person. "Jan would bake cornbread or muffins and we would have our staff meeting," Jack says, grinning. "All four of us sat around our kitchen table. Afterwards, our 'staff' would go back to their homes to continue SOLV work."

After five years, the small grassroots organization had amassed enough money to move its office out of Jack's home. The new location, however, had challenges. It was a condemned building in Hillsboro, unoccupied for fifteen years, with dry rot, uneven floors, and leaky ceilings and bathrooms. But it was SOLV's first real office.

The momentum, Jack recounts, only continued to grow. Part of the reason was the idea behind the movement was creative and unique: give the people of Oregon hundreds of opportunities a year to get involved in a lively spirit of volunteerism for something bigger than themselves. Provide people chances to participate at a time it works for them at a place near where they live. Let them help clean up Oregon's beaches, remove invasive species, and plant native trees.

"We knew not everybody can join a board, not everybody can donate five hours a week to a hospital. But they can be a part of 'episodic volunteerism,'" as Jack and Jan coined it. "There is a hunger in people; a spiritual, social, physiological need. People want to volunteer. They may not be able to attend the evening meeting, or volunteer every single Saturday. But they can take their children, or just themselves and their friends, and go down to clean up the beach or the illegal dump site, or paint out the racist or homophobic graffiti."

Jack describes SOLV's history with a genuine sense of pleasure. "SOLV became that barn builder. The old barn-raising organization for Oregon. The barn we were raising was the barn of community. It was the barn of coming together—the CEO being covered in the same dirt as the mailroom clerk. The managers for PGE, Tektronix, IBM with your forklift driver. People who you never have an opportunity to interact with. It was sharing experiences that were positive in nature and that showed immediate outcomes."

On SOLV days, weather didn't matter. Sometimes it was sunny, and everyone enjoyed planting native seedlings that would grow where once invasive plants had strangled the landscape. Or on a cold, rainy day, people worked together to pick up litter on a sandy, windswept beach—getting drenched in the process but smiling and sharing stories and feeling good about what they were doing.

The outcomes? Those beaches, once filthy, are now clean. The graffiti is removed. The illegal dump site is clean and the stream running clear again.

Jack and Jan, through eighteen years of effort, preserved Tom McCall's fourth legacy. With the help of Oregon businesses and eighty thousand volunteers each year working on over nine hundred statewide projects, the organization has grown from one employee in 1990 to a staff of twenty-eight and an annual budget of $2.6 million annually. Today SOLVE—now with an E —is firmly ensconced in Oregon values.

Its annual beach cleanup alone mobilizes over five thousand Oregonians to pick up tons of trash—broken pieces of Styrofoam, remnants of plastic cups and discarded plastic containers, car tires, cigarette butts, plastic bags and food wrappers—from project sites up and down Oregon's 362 miles of public beaches. Since 1984, over 130,000 volunteers have removed 1.7 million pounds of litter and marine debris.

Jack's passion to build community through volunteer action, has a deeper goal still. It remains his north star and still stokes the fire within him.

"There is an Oregon ethos. This promised land gave me so much . . . emotions I had never had before," says Jack. "Whether you're a fifth-generation Oregonian or a newcomer just unpacking the U-Haul, with your citizenship also comes a high degree of responsibility—both personal responsibility and responsibility to the Oregon community.

"The Oregon DNA is to become part of something more than just yourself. It is welcoming every Oregonian with: 'I don't care what your background is, what your political, your social, your moral stripe is; I don't give a damn.' Oregon is a place where there's the big table that everybody can sit around and look at one another . . . converse, not yell, discuss, not complain, not ridicule, and come away from that table saying, 'This land is good. There is goodness that defines this state. Let's embrace that goodness.'"

And if we don't? This is Jack's greatest concern.

"If we take Oregon for granted," he reflects gravely, "we are going to lose it. It is going to be like sand pouring through our fingers. We are going to wake up one day and say, 'What happened to our Oregon? When did it occur? Can we ever get it back?'"

Jack worries we are already heading down that road. He fears we don't hear that powerful cry, "This is our Oregon!" any longer. He acknowledges we no longer have former governors Tom McCall, Bob Straub, Barbara Roberts, or Vic Atiyeh to guide us. Thousands of people are moving to our state, but few have knowledge of the prior leadership that brought us to where we are now and made Oregon a place so attractive to live.

The rest of the country still looks at Oregon as progressive and desirable, Jack recognizes, but without understanding on a deeper level what it means to be an Oregonian, he despairs for his state's future. His ardent hope is that there will be those who hear and will respond to that clarion call.

"We need to create new meaningful, visionary ideas for this state, whether it manifests in legislation or just in the fundamental aspect of respect for one another. We need that direction, that person, that organization, that legislator yelling from the mountaintop, shaking us by the scruff of the neck.

"What's in our Oregon DNA?" asks Jack. "At its core is the knowledge that, where you are now, is a gift to you. It is the belief that this is one shared space for diverse people to find commonality. It is a deep hunger to give back what you've been given."

And the strongest flame? The one Jack finds unquenchable? He is unequivocal in his reply.

"To preserve this treasure called Oregon."

Parks are the most democratic institutions
in our great country, because they are
there for everyone.

———————————

Charles Jordan

2

Lifting as We Climb

CHARLES JORDAN

The first African American parks director in Portland history believed access to parks was not just a question of recreation, but of social justice. Thanks to his leadership, Portland's Forest Park—the largest urban natural sanctuary in the nation—is protected by a signature law designating the park as a city wilderness, where everyone, including those who cannot afford to travel far, can experience the grandeur of a true Northwest ecosystem.

∾

In November 2011, the City of Portland Parks and Recreation won the National Recreation and Park Association's National Gold Medal for Excellence. The achievement was a significant recognition, reflecting years of progressive thinking and actions taken by Portland Parks. Underlying much of that honor was the hard work and vision of a man named Charles Jordan, who for more than fifty years devoted his life to two causes he felt were of supreme importance and indivisibly linked: the welfare of people, predominantly children, and the environment.

The well-deserved award was just one of many Charles was involved with during his long career. Despite his many achievements, at seventy-three Charles reflects on his life with characteristic humility.

"Why me?" Charles Jordan asks, genuinely. "Of all people, why should I have been chosen to do this work? I did not aspire to any of these positions. But I did this work because I had to thank the people who believed in me more than I believed in myself. I could not let them down."

For Charles, it all began in the place of his earliest memories: Kilgore, Texas. In 1945, when he was eight years old, he and his family moved to the small, rural community in the south. As a young boy, Charles took the town's segregation for granted; it was just the way things were. But one very hot day, on his way home after picking up his mother's paycheck for her, he faced his first experience of direct racial prejudice, and it shocked him.

He had walked that day for an hour and a half on dusty, dry roads. He was thirsty. Spying a drinking fountain at a service station, he went for a drink of water. Immediately he was confronted by an angry white man who stopped him.

"You get away from there!" the man yelled at Charles. "You can't drink there!"

The memory, as well as other times when he faced discrimination, would sting throughout his life. "On the whole, however," Charles says thoughtfully, "within my own community, my remembrances are yet of a childhood sweet, sheltered, and cared for. Kilgore did not feel like a town, but like a village that cared. It was a hardworking community where I was loved, disciplined, instructed, and nourished among the forests and streams that provided food and my first connection to the earth."

It was also the place where Charles learned lessons he would carry with him his whole life: the worth of a place that values children and a deep attachment to the outdoors. Much of his instruction came from his mother, uncle, and his beloved Aunt Nannie, who was the one to first teach him the old spiritual hymn, "Lifting as We Climb." He never forgot those words, and with each achievement in his professional career, Charles strove to lift with him those who made up the mosaic of his life.

"I have always felt the urge to reach out and include others in any good fortune that came my way. It's just a lifelong practice of helping someone else every time I rose a step," says Charles.

He came away with something else too: a personal resolve that the hurtful racism he faced would never deter his views of humanity. He became determined to be "better, not bitter."

"My father believed that when people act badly, it's from fear and ignorance. We are all made in God's image and we carry that goodness underneath," says his son, Dion.

In 1950, when he was thirteen, Charles moved with his family to Palm Springs, California, where he lived on a Native American reservation outside of the city. He loved his new environs, discovered new friends, and . . . grew tall. Soon he was a teen standing six feet seven inches, and becoming noticed for his athletic skill. After graduating from Palm Springs High School, he was awarded a basketball scholarship to Gonzaga University in Spokane, Washington, and became one of only three African Americans at the school.

At the time, Charles's dreams were to make basketball his career, but a professor and Catholic priest stepped in to interrupt his plans. "Father Harrington was assigned to the sport of basketball and would travel with us everywhere we would go and play ball. At the end of my senior year, he took me aside. He said, 'I've been watching you for four years; people gravitate toward you. You'd be good in another field. With your talents, I think your career lays beyond basketball. I think a major in sociology would be a good fit for you. If you stay one more year to complete it, I'll see you receive a scholarship.' I wasn't sure what he meant, but I didn't want to disappoint him," says Charles. "So, I stayed that extra year."

After graduation, he returned to Palm Springs, got a job serving in the Parks and Recreation Department for the city, and became assistant to the city manager. He was the first African American to ever work at the Palm Springs City Hall.

Charles loved the work. He supervised two parks, three little league fields, and designed programs for the youth of Palm Springs. Married and happily employed, he had no plans to ever leave again.

What he was not expecting, however, was an out-of-the-blue call from a staff member from the office of the mayor of Portland, asking Charles if he might consider interviewing for the new Model Cities Program. The Model Cities Program, under the direction of President Lyndon Johnson, had been authorized on November 3, 1966, by the Demonstration Cities and Metropolitan Development Act, as part of the president's Great Society and War on Poverty initiatives. The program arose in reaction to concerns with unsuccessful urban renewal projects and urban violence.

Charles submitted an application but only, he admitted, because one of his closest friends who lived in Portland had asked him to. He subsequently forgot all about it. But when the the mayor's office called requesting he fly up for an interview, he decided to do it, as it gave him the opportunity to see his former Gonzaga teammate. Charles smiles as he remembers the interview.

"When I walked into the room, everyone stared. I didn't know why until later. A staff member told me, 'No one knew you were black!'"

Charles returned to his job in Palm Springs, never expecting to hear back. But several weeks later he received a call from a nice-sounding man who identified himself as Mayor Shrunk.

"The mayor personally asked me to come back to Portland for a second interview," Charles relates. "But what got me was the way he said it. He said, 'If you would come, we'd love to have you.' That's why I ended up taking the job as director of the Model Cities Program. I couldn't turn that down."

Charles, his wife, Esther, and his children, Trish and Dion, moved to Portland where he took up the challenging position as director of the new Model Cities Program in 1970. The scope of his new position was to develop new and alternative forms of antipoverty measures in municipalities. Charles took the charge seriously. He successfully enacted several programs with the intention of generating jobs, improving education, creating access to childcare, and reducing urban poverty. Always mindful of the importance of trees, he also took steps to beautify blighted Portland districts by planting the maple and gingko trees that now run the length of streets and boulevards.

While working with Model Cities, Charles continued to contemplate the value parks gave a city. He began to see public parks as a great equalizer, providing benefits to people whether rich or poor. That belief was the basis of many of the things he would later accomplish and became an unwavering conviction for the next four decades of his life.

"People of the city need to connect with the land. They need trees and they need parks," Charles says earnestly. "Parks are the most democratic institutions in our great country because they are there for everyone."

Enjoying the support of a strong board, Charles assembled people of color to work together for civic projects. He started senior services programs. He began nutritional programs for children and helped organize the

Community Care Association, which delivered groceries to families three days a week from the Bethel African Methodist Episcopal Church. Then came a bend in the road.

In 1974, the Model Cities Program ended. Somewhat providentially, at the same time there came a vacancy on the Portland City Council. A seated commissioner left for a new position with the Port of Portland. The mayor of Portland, Neil Goldschmidt, asked Charles to fill that vacancy. Charles accepted. Two years later, when the original commissioner decided he wanted to return to the council, Charles ran for office against him.

Charles won. With that victory in 1976, he became the first black commissioner ever elected in Portland. Even more, he attained the highest office any African American had ever held in Oregon's history—a past stained with a constitution that once contained an article barring blacks from residing within its state borders.

Charles is quick to say that he personally experienced very little racism in his job or in the city. Rather, "the community gave me support like I never expected." For Charles, his focus was "mission-driven," not merely "organization-driven." He felt a moral imperative to "fight for the resources to offer safety and support to children." His successes, ethical demeanor, and growing popularity contributed to his easy reelection to the Portland City Council in 1980. That year, he became commissioner of parks—which turned out to be his favorite role because it allowed him to concentrate on his conviction that the well-being of people was tied to the health of their environment.

"You can't measure in numbers the wide-ranging benefits of a strong, well-supported parks system. How many at-risk kids are safer because they have a place to go, people who care? What health costs are avoided because recreation helps to keep people fit? What businesses decide to move to an area because the quality of life includes an excellent parks system?"

Charles's skill in inspiring others garnered new attention and more funding for Portland Parks. After being reelected by a wide margin for a third term as commissioner in 1984, he intended to continue his work developing more recreation programs and expanding and preserving more parks in the city. That year, though, came another turning point in Charles's life that required him to carefully consider a major decision.

Not long after the election, Charles was contacted by a firm in Austin, Texas, which offered him a position as the director of parks in Austin. At the time, Austin was the fastest-growing city in the country. It was a job that would mean he could concentrate fully on those things that mattered the most to him—the development of parks and programs for kids and seniors. An added incentive was that he would be able to spend more time with his wife and young children, something he treasured.

After much deliberation, Charles accepted. He relocated his family to Texas and began work guiding a growing park and recreation department in a thriving city. He soon discovered the scope of his job offered other entries into environmental issues as well—opportunities he could never have imagined and that, in time, would put him on a national stage. The first, he relates, came as a surprising phone call.

"A woman I didn't know asked if I would meet her at her office adjacent to Town Lake, a park in the heart of Austin. I thought she was a citizen wanting to make a donation to the park, so I went there. When I arrived, who was waiting for me?" Charles grins at the memory. "Lady Bird Johnson!"

After recovering from his amazement, Charles sat down to listen to Lady Bird. She said she had heard him on the radio speaking about parks and his words motivated her. Lady Bird told Charles he was stating her dream, describing the direction she had always wished for parks and conservation.

That meeting was the beginning of a long and wonderful friendship. Charles and Lady Bird would meet often and spend their time "just visioning," as they called it. To their shared delight, much of the brainstorming came to fruition. One remembrance above all others continues to cheer him to this day.

"Lady Bird and I were both inducted in the National Parks Hall of Fame. It was a great honor. Even more, we were inaugurated at the same time!"

Later that year Charles was asked to serve on the Board of Trustees of the National Recreation and Parks Association (NRPA). He was also invited in 1985 by the office of President Reagan to join the President's Commission on American Outdoors (PCAO.) The latter opportunity was eye-opening for him. As Charles explains, the PCAO was tasked to investigate how people around the United States felt about parks and open spaces. To probe that question, Charles traveled across the country with thirteen other commissioners for a year. At each major city they visited, they asked Americans what "the great outdoors" meant to them.

For Charles, three things resulted that forever impacted his thinking and his future, and developed into a new passion for his advocacy. The first was being introduced to two people who would become close friends for the rest of his life: Gilbert Grosvenor, president of the National Geographic Society, and Patrick Noonan, president of the Conservation Fund. The second was a dramatic expansion of his view to the whole United States. He began to see the critical need for the preservation of public lands, for clean air and water, and for conserving the "wild places that feed the spirit." Lastly, Charles became acutely aware that there were virtually no "non-white faces" in the environmental movement.

"Thinking back to my childhood in Texas and California, and the black people who cared deeply for the land, I was determined to find a way to make the reconnection—both to show that people of color do have an abiding concern and to bring them to a more visible involvement."
But there was a deeper meaning for Charles. He understood that it was not only necessary to conserve wild places in rural areas, but also to approach wilderness preservation from an unlikely place, "from the very cores of our cities through the management of development and urban and rural planning." Saving natural and wild places within our cities was a question of social justice.

As Charles comprehended, those who didn't have the means to get out of urban areas to experience wilderness were left at a disadvantage. Few if any wild places existed for economically disadvantaged inner-city kids to experience and learn about the great outdoors—places to find the enjoyment, rejuvenation, and inspiration he had known as a child.

For the rest of his life, Charles fought for that cause. He worked to save parks and natural areas inside our cities for those who needed them the most. He reached out to talk about the importance of such places to all communities, even if not everyone agreed.

"If you are going to try to lead and do what you believe is right, sometimes you will have to stand alone. What people don't understand they will not value; what they will not value, they will not protect; and what they will not protect, they will lose."

In 1989, after leading Austin Parks and Recreation for nearly five years, Charles was contacted by Mike Lindberg, one of his closest friends and current city commissioner of Portland Parks. Lindberg's message was simple, but earnest. The City of Portland wanted Charles Jordan back. Lindberg said everyone wished he would return, take up where he left off, continue the work he had envisioned, and become the director of Portland Parks and Recreation.

Once more, Charles deliberated with his family. They knew, as did Charles, that in his heart, Portland was home. He was then fifty-two years old; this move would be his last. Portland was the place he loved, with friends he loved, and was a city he believed had within its soul what it took to be great.

Charles took the job of director of parks of Portland. He held the position for the next fourteen years, bringing life and stature to Portland Parks while leading it to new heights. Under his direction, Charles completed many of the goals he had set out to accomplish. He made sure that no child would ever be turned away from a park or park program because of a lack of ability to pay. He elevated the importance of keeping and expanding green spaces in cities, specifically, the value of preserving a "wilderness in a city" for citizens to experience—Portland's fifty-two-hundred-acre Forest Park.

The park was established in 1948 with the unusual goal of creating a natural sanctuary that had all the qualities of a wilderness area—a place for the preservation of wildlife, for the health of native plants, and for the protection of watersheds. It allowed a remarkable opportunity for all who resided in the city, including those who could not afford to travel to far-away places, to experience the splendor of natural beauty and wilderness. Charles's views were clear when he spoke to a packed audience at the City Club of Portland.

"You will not find any place in America, an urban park, so close to the heart of a city, that provides the wilderness experience five minutes from downtown. Forest Park is unique; it is priceless."

Making sure it would stay that way was a challenge, however. There were pressures from a rapidly expanding urban population and competing uses. Charles allocated monies to have the park studied by scientists and technical advisory committees to come up with a management plan to keep Forest Park an "unparalleled resource . . . and unspoiled naturalness of an urban wilderness environment." After two years, a comprehensive plan was completed, with explicit goals, unusual for a city park anywhere.

The plan placed the ecological health of the park first, requiring studies for the protection of its wildlife. It called for keeping the park a place for nature education, for passive use only, where users could find tranquility, safety, and spiritual renewal. Even more rare for a city park, the management document came before the Portland City Council and became law on March 10, 1995.

By the time Charles retired in 2003, Portland had achieved national visibility for its enlightened approach to parks. The *Chicago Tribune* wrote that Portland had "one of the most progressive park systems in the country." Portland Park's budget had increased from $27.2 million to $62.5 million. The department had grown from 283 employees to 408. Forty-four new parks and natural areas had been added, bringing the total to 228.

Through his tenure, Charles's belief that parks were our most democratic institutions stayed paramount. He never veered from his devotion to nature and children and the benefits of parks and recreation to society.

"Fun and games is about much more than fun and games," he wrote. "Where would our kids be if our recreation programs closed their doors? Home, studying? I don't think so. When the crime rate goes down, credit goes to law enforcement, but you and I should know better. What we have got to understand is that we will either spend money on trying to pull back kids who have gotten pregnant, who have joined gangs, who have become criminals—kids who have fallen into harm's way through neglect, through a lack of unconditional love and a lack of positive opportunities. Or we can spend money on recreation and urban parks, and we can give every kid a chance to be able to come home to his mother just once and say, 'Mom, today I was number one!' Our children need that and they deserve it."

Retirement did not stop Charles from working for the things he believed in. In 2003 his friend from PCAO days, Patrick Noonan, asked Charles if he would consider serving on the board of directors of the Conservation Fund—one of the nation's largest nonprofits dedicated to preserving parks and open space. The organization has preserved over 6 million acres of wildlife habitat, forests, green spaces, working farms, and historic sites. Charles agreed readily. Eventually he became chairman of the board of the Conservation Fund—and one of the first African Americans to lead a national conservation organization.

"When I was in my fifties, John Muir opened my eyes," says Charles, reflecting on what has given meaning to his life. "Reading his work, I began to see that, when I pick up a rock, that rock does not exist in isolation. It's connected to something else. And that something else is connected to something else."

There is an urgency in his voice as Charles continues. "We need to get this notion across to our young people. We need to help them understand that everything is connected so they won't be so quick to say, 'Well, this is *mine.*' If you can get people to start thinking that way—that we are connected—then we'll start acting that way. And if we start acting that way, we will be the powerhouse. We will be the nation we want to be."

Charles pauses, then says what is in his heart.

"I am seventy-three years old and battling Alzheimer's. I'm doing well; what's more, I'm satisfied. My only regret is that there is so much that I want to do. I had no idea I would get to the level I am around the country, with all of these people," he adds modestly. "But God did. And now that I look back on it, I see it made a big difference. I had more confidence than I realized."

He reaches deeper inside himself, then continues. "You never think you have enough. But you just have to tell yourself, yes, you have the courage to make that step out there. What I have learned is that, now and then, God lets you know that he is with you—otherwise you couldn't do what you are doing. He is with you, so don't give up! When the times get hard; don't quit! You have the assurance that He's watching over you."

Taking another moment to reflect, he smiles, the kind, generous smile that all who love this man have come to know. Even the scourge of Alzheimer's can't contain it; his wisdom still can penetrate the darkest shadow.

"Do you want to know why I refer to all people as my sisters and my brothers? It's because we are all family. And the reason things aren't working well is that we don't think of each other as family. We think instead, 'I can be all of this; I can deal with all of this myself,' but in truth, we can only do it if we function as a family.

"We've got to be family. There are no winners or losers in this. We need to work together. We need to look out for one another. Don't you see? You and I cannot run this life to just try to get it all for ourselves, just to get out ahead of somebody else. No! We've got to do it together. There are some of us who are not as strong as the rest. Leaving them behind is not going to do it. In all we do, we've got to be family. In all we do, we need to think, how would you treat your family?"

As his son, Dion, expresses, his father, having been excluded as a child in a segregated community, worked throughout his life tirelessly to include others. He learned lessons of faith from his father, "this great, tall man who knelt down to pray every night. My father knows in the eyes of God he's small—and that governs his decision making, and it's why he speaks to everyone on the street—with respect—because each person is *somebody*."

The life of Charles Jordan is indeed a testament of someone who believed in family. From his great love for people and the outdoors, this caring man left lasting imprints on Portland and its park system. And always, while he climbed, he never forgot to bring others with him. Facing an uncertain medical future, it is his time to be held up, and Charles is not afraid.

"God has brought me this far," says Charles with perfect clarity, "and has never left me. So what makes me think that he's going to drop me now?"

If you're seeing, not just looking, then when you observe that tree growing in the forest, or the water bubbling up from the ground, or the sanderling on the shore, you are getting connected. That is what makes the difference. You begin to see yourself *in* a context rather than think you *are* the context.

———————

Neal Maine

3

Pacific Light
NEAL MAINE

Since 1986, the North Coast Land Conservancy has preserved nearly five thousand acres of the Northern Oregon Coast. Today it is close to achieving its greatest dream—a park that would stretch from the top of the Coast Range all the way to the sea. The Rainforest Reserve will protect thirty-five hundred acres of mountains, forests, and shore, conserving an entire watershed and extending three miles into a marine reserve. It will become an extraordinary natural park, "a mountain range precipitating itself into the sea."

∾

He thought it was the largest channel catfish ever hooked from the Big Blue River near his hometown—Blue Springs, Nebraska.

It turned out to be a tire.

"But not just any tire; it was a perfect fit to our old family car, a Model A Ford," says Neal Maine, recounting his father's angling experience in 1944. "And a great stroke of luck."

It meant that the family now had the necessary number of tires allowing Neal's father, mother, brother, and sister to make the trek to Oregon.

"Why my father thought he needed nine tires, I'm not sure, except for the fact you couldn't buy a tire in the middle of the war," continues Neal. "The one my father fished out of the creek was remarkably the right size. My family migrated west, along with half of Nebraska who was broke, still suffering from the effects of the Great Depression. There was a rumor there was work on the West Coast. That tire put us over the hump. We left with $40, a tank of gas, and everything we had in that Model A, heading for Oregon."

After a few stints as a contract worker in logging camps, Neal's father got a job on the Oregon Coast. Homes in Gearhart, Oregon, had become buried in sand as a result of the building of the South Jetty near Astoria. Neal's father's job was to save the houses and to raise them up, out of the sand. The job provided steady work. He ended up locating his family in Seaside, where Neal's aunt and uncle lived. At the time, Neal was six years old.

The Oregon shore enthralled Neal. He loved the clarity of the air, the Pacific light, the sparkle of the sun on fir boughs after rain, the misty morning fog. He roamed the windswept, sandy shorelines, the coastal headlands, the old-growth forests, and the salty estuaries. He became captivated with the birds and wildlife that inhabited the quiet spaces. Before long, the northern Oregon coast had set an anchor deep within his soul.

While he could not know it then, he would never leave the community for long. It would become his lifework to lead others to pay attention to the wonders that he saw all around him. In time, he would help save thousands of acres of the native, coastal landscape for the benefit of all future generations.

Neal's first foray of any length away from the Oregon coast occurred when he was eighteen years old and headed to Linfield College, a small liberal arts college in McMinnville, Oregon. Ironically, during his time there, he found another element mooring him to the shore. Karen Boyer immediately caught his eye; when they spoke, he was surprised to discover they had both attended the same high school in Seaside. Somehow they had never noticed one another.

"Karen grew up in Cannon Beach, I grew up in Gearhart. In that era, it was a million miles away," says Neal. "We married in our junior year. After graduation, we decided to return to the coast, where I landed a job teaching at Seaside High School—the same high school we'd both attended. I planned on staying only a couple of years."

Two years of teaching went by. Then, two more. Eventually, the years stretched to thirty-one, as Neal discovered that teaching was work he deeply loved and he greatly enjoyed young people. He found them honest, sincere, and eager to learn.

Neal quickly discerned, though, that the institutional structure of high school classes nearly snuffed out any flame of real interest in biology. Students' goals were directed to only one thing: "getting test answers right." The four-hundred-page textbooks were written in Texas and had scant relationship to the amazing natural world surrounding the school. Neal loved biology, his major in college. And only twenty-three steps away from the school lay a rich, coastal estuary full of living, teachable examples. Yet his students were bored.

"One day I said to myself, I'm not doing this anymore! So, I pushed my desk in the wastebasket, so to speak. I went down to the shop, taking the biology books with me. I used the school's band saw to cut the backs off all the books. Then I filed all the pages."

The textbook was no longer a memorization strategy. Instead, Neal assigned only those pages that dealt with real world studies, to be used as a reference for the community science they were doing. It worked. Curiosity was ignited.

Soon Neal's students were studying snowy plovers in the wild, researching habitat, making videos, conducting wildlife surveys, designing water systems, creating health fairs for the town with over two hundred people attending. With the help of fellow teacher Mike Brown, the two teachers originated a nonprofit inside the high school: the Coastal Studies and Technology Center. Educators from other districts began partnering with the for regional studies. Science studies at Seaside High were becoming a community resource.

The program Neal created would evolve to become a new way of learning. Highly innovative, it was a sweeping approach to engage students to design authentic work to study the landscape around them and their context within it. What Neal wanted his students to observe, and to carry with them, was the founding principle of biology.

"Everything is connected. Whatever we do, we don't do it alone. You can't undo twenty million years of nature making the rules," he explains.

In 1979, after twenty years in the classroom, Neal took a sabbatical leave from teaching and traveled throughout the country on behalf of the Annenberg Foundation. During this time, he instituted this new paradigm of "place-based education" in multiple schools. Neal helped teachers across the United States design research projects that would contribute to the larger community around them. When he returned home to Seaside High School, he continued to expand science programs, bringing five or six neighboring schools together for increasingly sophisticated research projects.

These shared investigations, co-directed by Neal and Mike Brown, were immensely successful. One particularly exceptional project even attracted national attention. Partnering with National Marine Fisheries, students conducted research on the invertebrates living in the shoreline benthic zones of the Columbia River jetty. They won a national Environmental Protection Agency award for using science to contribute to communities. "What's more," Neal says, "the kids got to make a presentation on the White House lawn to President Bill Clinton. They all shook the president's hand and received their award from him in front of the White House!"

While deeply devoted to his profession, Neal also became involved with working to protect Oregon's natural areas in other ways. In the late 1970s and early 1980s, he volunteered his time to Oregon Shores Conservation Coalition, still active today, defending the state's coastline through natural history education and preservation action. He also sat on the advisory board for Senate Bill 100, Oregon's land-use law. The landmark law, enacted on May 29, 1973, was the first of its kind in the nation. It created a foundational structure for land-use planning goals throughout the state of Oregon. Senate Bill 100 would act as a springboard for much of the work that Neal would be able to accomplish in the future.

"With its nineteen goals that all planners need to consider, Senate Bill 100 is an incredible resource. It has allowed Oregon to be an island that doesn't give up everything to sell out to the highest bidder," says Neal. "The law for comprehensive land-use planning leaves no doubt about its mission. Every city council? They know about it. They have to plan. Senate Bill 100 gives all cities and counties in Oregon a base that's already established. You don't have to make up a new one each time, which is where a lot of things get sidetracked."

As Neal worked to protect important properties on the Oregon Coast, though, one thing continued to plague him. There was, for all purposes, only one strategy environmentalists used to save critical pieces of the landscape. "It was called," Neal explains, "take the developers to court. That was a tough show. The people you were up against had a lot more time and a lot more money. The projects were all challenging. Everybody was exhausted and we all were beginning to feel this isn't going to be the answer."

In 1985, however, the San Francisco–based Trust for Public Land—a nonprofit organization created in 1972 with the goal of creating parks, protecting land, and connecting communities to the outdoors—began reaching out to smaller communities in Oregon and Washington. Neal saw right away it could be a new ally in the land-use battles. The organization's objective was to encourage local residents to create regional land trusts to help preserve important natural sites in their own areas.

An attorney for the conservation group held meetings up and down the west coast to explain the idea behind land trusts: it was a cooperative approach to conserving key properties. Rather than just filing lawsuit after lawsuit, the land trust philosophy was to encourage environmentalists to work with landowners willing to preserve important pieces of native habitat. For significant properties, land trusts worked to purchase the lands outright or use the tool of acquiring conservation easements. Conservation easements are legally binding agreements between a landowner and a qualified private organization or public agency in which the landowner sells or donates certain rights to their property, usually the right to develop or subdivide it. The landowner continues to own and manage the land. The agreement is in perpetuity, continuing if the property is later passed on to heirs or sold.

Neal attended one of those meetings. That night, the idea of creating such a conservation trust for the northern Oregon Coast took over his imagination and that of many others present. The more he contemplated it, the more the idea wouldn't let him rest. He began talking with others who were interested in this new method for protecting the coast. Before long, the spark had grown.

One year later, in 1986, Neal helped found the first land trust solely for the northern Oregon Coast—the North Coast Land Conservancy. An eight-member board of directors wrote the charter, creating a nonprofit organization in 1987. Neal became its executive director. When he retired from teaching in 1993, Neal took that role full-time and remained in the position for the next fifteen years.

"The North Coast Land Conservancy began with a couple of little projects, but nothing of real significance happened until we had a crisis situation—one that took us by complete surprise," Neal recounts. "None of us knew about the threat to Saddle Mountain State Park, that is, until loggers were literally coming over the hill at us. They said they were planning to cut right up to the edge of a beautiful old-growth forest inside the state park—the last remaining stand of old-growth Douglas fir in Clatsop County. Before we saw them, we could hear the chain saws."

As Neal learned, the forest was slated to be cut in a period of weeks. It directly abutted the prized old-growth section—a place recognized for its high biodiversity, rare plant species, and monstrous Douglas firs—and would destroy the buffer between the clear cut and old-growth ecosystem.

"When you cut right to the edge of an old-growth forest, you lose anywhere from six hundred to one thousand feet of old-growth habitat because of the temperature, light, and variables that go right into the forest," Neal explains. Preserving the critical barrier, which included an integral part of the Fox Creek Watershed, suddenly became of paramount importance to the newly founded North Coast Land Conservancy.

In the past, the only course of action would have been to bring a lawsuit, if that was even possible. The North Coast Land Conservancy attempted something different. In the course of a few days, Neal and members of the board met with representatives of the logging company that owned the buffer property, Cavenham Forest Industries.

"We asked them, 'What kind of deal would work for you? We want to try to work out something that is cooperative.'"

The tactic bought them some time. Neal and fellow board member Doug Ray were successful in convincing the company to put a temporary stay on the cutting. He told the representatives he would endeavor to find an alternate place for them to cut—one with habitat not as critical, but with the same land base, the same acreage—in exchange for keeping the buffer intact. Eventually, North Coast Land Conservancy located a piece of landlocked property in the middle of another forest owned by state parks. The timber was second growth, not as ecologically significant, and already scheduled to be re-harvested at a future time.

Cavenham agreed to the swap and the buffer was preserved. The transaction, coordinated by the land trust, became the first biodiversity-based conservation proposal ever pursued by local activists involved in the negotiations.

As it turned out, the agreement was even more of a win-win, according to Neal. While the North Coast Land Conservancy was working on the exchange, two endangered species, both listed, were discovered on the property: the marbled murrelet and the Cope's giant salamander. This alone could have been used to take the timber company to court.

"We chose not to," says Neal. "Instead, we didn't even tell them about the discovery until the arrangement was eventually cut. Then we announced that there were endangered species in the buffer, which made their contribution an even bigger deal!"

That experience—working to try to find some common ground between willing partners rather than acting as adversaries—set the pattern for the North Coast Land Conservancy. It would seal the land trust's innovative framework for conservation.

"Our organization is not looked at as an environmental group; rather, it's part of the community resource. It's about helping people realize the community is the baseline; everything rolls from that. We don't live in isolation; we are part of a larger structure."

After the victory in saving the integrity of Saddle Mountain State Park, the North Coast Land Conservancy took on many more projects, and its reach has grown exponentially. Its focus, though, remains consistent: utilize strategic planning to maintain natural waterways, corridors, estuaries, forests, and wetlands on the northern Oregon Coast.

"Because of Oregon's strong land-use laws that have teeth and validation, our land trust has been actually able to be the good guy, coming and helping communities, private landowners, and even developers find ways to be successful. We aid community planning, where we can slip in and protect the critical resources, our mission paralleling the goals of Senate Bill 100."

Today, thirty-two years since its founding, the North Coast Land Conservancy has protected over fifty properties valued at millions of dollars, parcels stretching from Astoria to Lincoln City and totaling close to five thousand acres. The majority of its holdings it owns outright. Further, the North Coast Land Conservancy cares for each one itself, and every property it owns has its own management plan. People sign up to volunteer as stewards for the sites, continually checking on them and turning in their notes to the conservancy's stewardship director.

"Most of our properties can't be duplicated. Size and scale is different on the coast, in comparison to other areas," says Neal. "We have projects and own properties that are only five acres, but the right five acres can be essential. In everything, we look at a connectivity concept, focusing on a systems level. Connections are arteries and veins. You have to have systems that are connected or conservation doesn't work."

Perhaps the most ambitious project ever undertaken by the North Coast Land Conservancy is on its way to becoming a reality: the Rainforest Reserve. Neal beams as he explains it.

In November 2016, the conservancy signed an agreement to acquire 3,500 acres of coastal forest. By 2020, it hopes to raise $10 million to finalize the impressive dream. When completed, the land will connect Oswald West State Park, bordering the ocean, to Onion Peak, a mountain in the Coast Range of Oregon, east of Arch Cape.

The Rainforest Reserve is an enduring vision with roots reaching back three-quarters of a century. The concept was originally imagined by Samuel Boardman, Oregon's first parks superintendent from 1931 to 1950. Affectionately known as the "father of the Oregon parks system," Boardman made it his personal and professional mission to acquire and preserve significant lands in Oregon for parks. During his nineteen-year tenure, park systems throughout Oregon grew from 4,000 to 66,000 acres.

An important part of Boardman's legacy was the protection of 354 acres in Oswald West State Park, which today has grown to 2,448 acres. The park, located on the north Oregon coast, is a fitting tribute to its namesake, Oswald West. As Oregon's fourteenth governor, West established the state's first major protection of Oregon beaches in 1913, declaring they remain as "public highway" with public access to the high-tide line.

When Oswald West Park was dedicated in 1931, Boardman dreamed that its boundaries would grow, that lands extending from the state park would be protected all the way to the crest of the Coast Range. He strongly believed that, with this connection, it would become a crown jewel in Oregon's park system, one of the "outstanding natural parks in the nation. . . . In few places in the nation do you find a mountain range precipitating itself into the ocean."

Boardman passed away in 1953, never living to see his vision come to pass. His farsightedness, though, as Neal relates, was not forgotten.

In 2006, the North Coast Land Conservancy resurrected Boardman's idea. The Rainforest Reserve became one of the conservancy's major challenges. Now, thanks to the conservancy's steadfast dedication, Boardman's conception has at last the chance of coming true.

"It's got it all!" Neal exclaims. "When finalized, an entire coastal watershed covering twenty-nine square miles—from the headwaters in the mountains to the Pacific Ocean—will be protected forever. Making it even better, this corridor includes Cape Falcon Marine Reserve, which extends three miles offshore. The venture will be like none other in the United States!"

Neal stepped down in 2008 as executive director of the conservancy, passing that torch to Katie Voelke, who continues to tirelessly work to see the Rainforest Reserve come to fruition. After serving for two years as the organization's conservation director, Neal officially retired. For the last ten years he has been pursuing a new adventure: photography with a purpose.

"After I left the front lines of the land trust when I turned seventy, I had a conversion of sorts," says Neal. "It's still conservation, but it's now centered on an individual basis. It's about quality of life. Growing bigger is not necessarily better. In relationship to quality of life, bigger can become more harried, more controversial, more 'everything' if it's just bigger.

"It's not the quantity, but the integrity and the quality of life that count," says Neal thoughtfully. "I have seen how, if you are paying attention, the quality of your life goes up. Way up! I hope to lead people through this idea, the one captured by Thoreau: 'It's not what you're looking at that matters, it's what you're seeing.'"

To try to increase his own awareness, Neal began accompanying artists and photographers outdoors, asking them, "'Show me what you're seeing.' I wanted to try to understand how these artists collectively pull together all they are seeing, how they are trying to send a message about the world through their art."

Through his artistic explorations, Neal developed a new attentiveness to all the things he loved about the coast. He spent long days taking photographs of its wildlife, estuaries, forest glades, and streams. He formed a partnership with his grandson, Michael Wing, an exceptional photographer, and created Pacific Light Images. Like many of the artists he has learned from, Neal's own art has a message. By showcasing the richness and variety of the coastal landscape with striking photographs, he endeavors to raise people's mindfulness of the ecology of the Oregon Coast and to lead them to recognize its true beauty and value. He calls their work "The Art of Seeing."

"If you're seeing, not just looking, then when you observe that tree growing in the forest, or the water bubbling up from the ground, or the sanderling on the shore, you are getting connected. That is what makes the difference, and the effort behind our art. That is when you begin to see yourself *in* a context rather than think you *are* the context."

Neal is clear that protecting natural areas is not an easy task. He admits that, for all our labors, we may not always know the outcome, even in our lifetimes. One has only to consider Sam Boardman, who never quit advocating for the creation of a "crown jewel" of a park, today known as the Rainforest Reserve. But the rewards for trying, Neal says, are worth more. Much more.

In over a half century of pursuing conservation, Neal has seen ventures come together in ways he could never have imagined. While it has taken perseverance and devotion, he has discovered a principle that, Neal holds, underscores everything:

Stay with it.

"Through all my years with the North Coast Land Conservancy and beyond I have come to realize that is the important thing," Neal imparts. "Stay with it! And believe, if you are doing the right thing, it will, ultimately, turn out right."

The project was too important to give up.
I knew the canyon and I loved it. That is what
gives you the courage to go on. Even when things
got hard, I could always picture the whole thing.
Through it all, I could visualize the pleasure
it would be for everybody.

Lucille Beck

4

The Kitchen Counter Drive

LUCILLE BECK

Dazzling native trilliums are emerging once again in Tryon Creek State Park, clothing the natural area's hillsides in springtime white. Today, the park is celebrating the return of the flower with its annual Trillium Festival, enjoyed by families throughout Portland and the region. Naturalists are leading guided hikes; interpretive volunteers teach about wildlife, soils, pollinators and trees at science stations. Hundreds of people follow the pathways to enjoy the beautiful displays of spring wildflowers, ferns, and Oregon grape. What most of them don't realize is that without the devotion of a local housewife, the park would be covered over with condominiums.

∾

For months, the Multnomah County Commissioners knew that two hundred acres of pristine, forested canyon had been optioned for sale to a developer. They were aware, and supported fully, the plans quietly being drawn for a massive complex of condominiums, apartments, and a tennis club that would transform the natural ravine with the sonorous stream known as Tryon Creek running through it.

They just didn't tell anyone.

In particular, they didn't enlighten Lu Beck when she and a small group she'd formed came to them in 1970 with an altruistic dream to create a county park in this landscape adjacent to the city of Lake Oswego and five miles south of Portland.

"They encouraged us, even complimented us on the little brochure we'd made to help garner interest in creating a park!" Lu recollects, still with a hint of disbelief. "They disclosed nothing about what was already in motion—that hundreds of acres in the canyon were on a fast-track for development; even more, that this was the option they stood behind."

Lu pauses, remembering, shaking her head. "As we would come to realize, from the start nearly all the commissioners had no intention of ever creating a park."

As Lu was to discover, developing two hundred acres of the canyon was just the beginning. The county's true vision was far grander. The Multnomah County commissioners imagined the entire 657-acre forest known as Tryon Creek Canyon becoming a series of massive subdivisions.

"Perhaps it was a good thing we didn't know the immensity of the county's conception right away," Lu says with a sigh, "or else our task to protect it might have seemed formidable if not impossible!"

Lucille Smith, known to everyone as Lu, was born in Portland in 1925. Her childhood was not an idyllic, worry-free existence. When she was ten, Lu's father died suddenly. To keep the family intact, her mother, Helen Brayton Smith, took over the family business, Smith Auto Parts, while raising two pre-adolescents in the midst of the Depression. As Lu relates, "It was quite unusual for a woman to be running a business at all, let alone an auto parts business. It wasn't easy, but my mother had good help, persevered, and somehow kept the company alive for another twenty-five years."

Lu always had a strong love of nature, developed from her years in Camp Fire Girls, spending summers at Camp Namanu on the banks of the Sandy River, east of Portland. "I went to the Camp Fire camp as a camper and later, in high school and college, as a counselor. It hugely enriched my life. Maybe that's a little unusual. But that was how I came to love nature— being outside and hiking and loving the rushing brooks and streams."

In 1954, Lu married Borden Beck. He was a young attorney and another Oregon native who hailed from Redmond, in the central part of the state. Borden was an avid outdoorsman, and growing up spent all his free time hunting and fishing in the desert canyons along the Deschutes and Crooked Rivers. The young couple soon bought a country home in Multnomah County, north of Lake Oswego and situated on the rim of Tryon Canyon, at the time a large, undeveloped wooded area.

As the cities of Portland and Lake Oswego developed over the next fifteen years, the forest remained amazingly untouched. Before long, she had fallen in love with Tryon Creek Canyon; each year its beauty and character only captivated her more. The sylvan ravine boasted carpets of trilliums blooming in spring. In fall, the woodland dazzled with color—bright yellows cast from big-leaf maple leaves and crimson reds from native vine maples. The clear, glimmering creek—a freshwater stream and tributary to the Willamette River—stitched the canyon together, winding through the quiet forest.

Lu loved to walk along the hillsides, enjoying the water with its steelhead trout and spawning coho salmon and relishing the quiet solitude. It seemed natural to her that this healthy watershed—an oasis juxtaposed between two rapidly expanding cities—should be considered for a park.

In 1968, she and her husband discussed the idea with two neighbors, G. H. Oberterffer ("Obie") and Nancy Park, who agreed they felt the same way. The four scheduled a meeting with Multnomah County officials. There, one commissioner, David Eccles, appeared to have some interest in establishing at least a small park in the canyon and divulged that the county was considering an option. Lu and her friends were delighted when, a year later, Multnomah County purchased forty-five acres in the Tryon ravine. Lu took that as affirmation that the county would be on board for creating an even larger park.

Her excitement kindled, Lu felt now was the time to begin enlisting broader public support for the idea. She reached out to people and arranged a neighborhood meeting in Lake Oswego; over forty citizens showed up. And then, fortune smiled on her. On one of her daily walks, she met someone whose passion for Tryon Creek rivaled her own.

Jean Siddall was an amateur botanist who delighted in the richness of the native plants of the woodland. As Lu passed by her on a forest trail, the two paused to talk, discovering they were neighbors. Jean told Lu that she had explored the canyon for years, roaming it throughout, and declared it was amazing. It showcased a veritable wonderland of plants—native trees, shrubs, and wild flowers—a prize so close to a metropolitan area. Each time she visited the forest she became only more cognizant of its value. Few streams in Portland could claim native trout and salmon. The native birds that cheered the woods with their songs depended on the natural habitat Tryon Creek provided.

With such similar interests, Lu and Jean quickly became friends. Before long, they shared the goal to do all they could to protect the treasure of Tryon Creek Canyon. The first thing, they decided, was the need to seek out other like-minded individuals who were willing to truly commit themselves to the task. Nancy Park and G. H. Oberterffer, Lu knew, were two. Lu reached out to another neighbor, Betsy Dana. Soon, more neighbors were enlisted.

"Seeing that we now had a little clutch of people who were strongly willing to help, I felt emboldened enough to go to the county commission to present our idea to create a larger park," says Lu. "We made an appointment and spoke to all the commissioners. At the meeting, they encouraged us, and seemed very receptive. With their backing, I felt we were on our way."

What Lu would find out later was that at the same time she was entreating the commissioners for a park, a major property owner, Ward Cook, had presented his own proposal to develop Tryon Creek Canyon. While placating her small group for their advocacy to create a park, the county commissioners were far more enthusiastic over Mr. Cook's plan. This was the option they were giving their full backing.

"From the start, the county never had any intention in creating a park," says Lu, shaking her head. "Mr. Cook had retained a developer in Seattle and had a blueprint in hand for a housing construction project spanning over two hundred acres of Tryon Creek. I quickly saw that protecting and preserving the beautiful canyon was going to be a much harder endeavor than I had thought."

Facing an uphill battle, Lu took stock of what she had. "I had the full support of Borden, my husband, who never once wavered from the idea that the canyon below us should be set aside as a park. I reminded myself, there were lots of people who felt the same way. I just needed to collect them all together somehow, and get the whole thing started."

Setting out to do just that, Lu brought together a core group—Jean, Nancy, Betsy, and herself—to determine a course of action. They decided they needed to let the county commissioners know that there was direct opposition to Ward Cook's plan to develop the canyon. Then they needed to begin gathering public support. Lots of it. They also recognized that if a Tryon Creek Park were ever to happen, they needed to raise money to attempt to purchase the property from the developers. It was obvious Ward Cook would never abandon his development plan for altruistic reasons. Therefore, they would need money and community backing to have a chance.

Lu's thinking didn't stop with Ward Cook and his two hundred acres, though. She knew they would be spending hours of time and energy to try to stop this one development. She thought, why stop there? What about a larger objective? Soon the project had taken on a new mission, one Lu enthusiastically embraced.

They would expand their goal. They would work to save the entire 659-acre canyon.

Lu dove in. First, she called Mr. Cook's developer in Seattle. When he answered, she was surprised that he was already aware of who she was.

"Mr. Jones, the developer, was cordial and immediately tried to reassure me that what they were proposing would be a great benefit to the area," she remembers. "He said their development would increase property values of all our homes. He said that the condominiums and apartments would be very high quality, there would be a wonderful tennis club. Then, he added for my sake, 'and you will have some park!'"

Lu rolls her eyes. "I thought, sure we will."

She decided that they had to go public. "I knew we weren't going to get any help from the county. That meant we needed to tell everyone we knew about the benefits of creating a park. And so we came up with a plan." Lu needed volunteers—many of them—to spread the word. She formed a small board of directors to figure out how to proceed. The first item was what to call themselves. It didn't take long to come up with a name: Friends of Tryon Creek Park.

The next step was to elect a chief fundraiser. Barbara Ehrman, an active member of the League of Women Voters in Lake Oswego, volunteered. Her office had a mimeograph machine. She would start printing flyers and newsletters.

In search of recruits, Lu attended every neighborhood meeting she could find. At these events, she exhorted all those who lived near the canyon to reach out to other neighbors and to begin to fundraise to buy options on properties before more were sold to developers. She told of the incredible beauty of the canyon. She described the colorful red-flowering currants and pink salmonberry arching over the stream, the graceful sword ferns growing in profusion in the shadowy glades. She explained the benefits a wonderful park would bring to the community and to all who would visit it for years to come.

Lu's speeches were well received and many volunteers joined the cause. When the initial five presentations were over, Lu had done something that the county never anticipated. She had changed a message into a movement.

Moreover, it would become a crusade unlike anything the County had ever witnessed before. The volunteers, as Lu relates, were all women, "310 housewives taking time out of their busy days to participate in a fund drive in the manner of a 'mini-United Way.'" Surprised by this entirely unexpected show of force, the county wasn't quite sure how to deal with it.

"The commissioners came up with a term for us, one that we gladly embraced," Lu remembers, smiling. "We became known as the Kitchen Counter Drive."

With the first Earth Day approaching that April 1970, Lu felt the timing was perfect to enlarge the mission. The newly formed Friends of Tryon Creek Park, which had received nonprofit status, would capitalize on the upwelling of energy. Schools were teaching classes about environmental issues. Well, wasn't saving Tryon Creek an environmental issue?

Barbara Ehrman went to work developing a presentation, brochure, and a kit with handouts. Lu and Jean began speaking at schools and garden clubs and community groups. They brought with them something else that perhaps only a Kitchen Counter Drive could come up with.

Lu grins. "Buttons."

While not exactly a blockbuster fundraiser, the buttons resulted in a dramatic increase of the issue's visibility. They were bright yellow, emblazoned with the words "Tryon Creek Park," and sold for twenty-five cents each. Soon, everyone was wearing a button.

The increasing involvement of children, neighbors, civic groups, and schools caught the attention of the press. *Lake Oswego Review* staff writer Frances Davison loved the idea of a new park and took on promoting it as her personal directive. Each week she wrote local-interest stories with pictures, creating more visibility. If a school hosted a bake sale to raise money for Tryon Creek Park, Frances noted it in the *Lake Oswego Review*. If there was a benefit at a school dance, it made it into the paper. Numerous cake walks and spaghetti dinners, all raising money for the park, were written up in the *Review*.

Lu knew, though, that while well-intentioned, all the community fundraising would not go far enough to buy options on land in the canyon, or even to entice Ward Cook to sell his property and give up his dream of developing it. They needed more publicity, to reach out farther afield, to get more people involved. Lu came up with the idea one evening; it was a strategic stroke of brilliance.

Selling trails.

Of course, everyone knew there were no trails yet in Tryon Canyon, nor proposals to build any. But to Lu, the idea was the visualization that someday there would be a park that encompassed the canyon, with hiking trails carefully situated where people wandered to see the glorious diversity of wildflowers and resident and migratory birds.

The Friends of Tryon Creek Park began selling trail certificates. Each $10 donation purchased one foot of trail. When someone contributed, they got a certificate for however many feet of trails they had procured. It became a receipt and a memento. People started buying. Families purchased not one, but two, three, or even four trail certificates—one for each child—so when the park came to pass, they would know they had played a part in it. Soon the Friends had collected $27,000 from fourteen hundred families.

"Everything was for this wonderful park!" says Lu. Then the shoe dropped.

Early one morning, Lu received a surprising phone call from Mr. Jones, the developer in Seattle. He said he wished to consult with her and offered to come to Portland to talk. At the meeting the following week, which they held at her husband's office, Mr. Jones came straight to the point.

"I must say your efforts are commendable. We all know, however, even with all your contributions, you'll never have enough money to buy the two hundred acres," said Mr. Jones. "Even if you could, it would make little difference. You see, we are not only planning to develop the two hundred acres. We are going to develop it all, the whole canyon," said Mr. Jones.

Lu sat back, astounded. Mr. Jones continued, "And we will create a nice little park in the middle."

"How do you plan to get all these properties, Mr. Jones?" she asked. Mr. Jones took in her stricken face and said, enigmatically, "It's all been worked out."

It was perfectly clear to Lu that if there were any hope of creating a Tryon Creek Park, they needed to stop Mr. Jones and Mr. Cook from going ahead with their campaigns. They also needed to stop any other developers from acquiring properties in the canyon. The question, of course, was how?

After the meeting with Mr. Jones, they faced their first challenge. Jean noticed a "For Sale" sign erected for a thirteen-acre property in the canyon, the first of many lot sales threatening Tryon Creek's preservation. Lu set up a meeting with the owner of the property and asked him what terms he might consider. The landowner was unwavering in his demands. He wanted $39,000 for the thirteen acres, with $25,000 down and the rest on contract. There were others, he told Lu unflinchingly, who expressed interest in buying his land.

"I went home and immediately called Jean, Nancy, and Betsy, our charter members. There was little time to act," says Lu. "We needed to find five people to each put down $5,000 right away, or twenty-five to donate $1,000."

That afternoon Lu received a phone call from someone whose voice she did not recognize. "I understand that I'm buying some property out on Tryon Creek?" he asked.

"You are?" asked Lu, incredulous.

The man identified himself as Betsy Dana's father. He was a forester and the director of the Society for the Protection of New Hampshire's Forests, a private, nonprofit land conservation and sustainable forestry organization based in New Hampshire. Established in 2001, the association presently owns more than 180 properties covering 54,000 acres. He recounted he had visited his daughter the previous summer and had hiked through parts of Tryon Canyon.

"Without any hesitation, he said he would send us a loan made out to the Friends of Tryon Creek Park for $25,000 so we could buy the first thirteen acres outright," says Lu, with some amazement still in her voice. "So, we went to work right away. We got a contract from the seller with the rest to be payable over ten years." Lu's face gleams. "It was the first thirteen acres of Tryon Creek Park! Now we had only 646 acres left to go!"

When the second property came on the market—twenty-eight acres not far from the first site—Lu convinced the owner to give them an option on the land. Ultimately, he agreed. Then the Kitchen Counter Drive began contacting all of the owners of the real estate in the canyon, with the same urgent message: "We are from the Friends of Tryon Creek Park. You may be hearing from a developer in Seattle about wanting to buy your property. If you do hear from him, please don't sell him your property! We would greatly appreciate your talking with us first!"

Lu decided now was the time to approach Mr. Cook himself.

"We had no idea where the money would come from," she says with a laugh, "but Jean and I met with Mr. Cook and offered him $850,000 to purchase his property. Incidentally, that property is the same spot where Tryon Creek's Visitor's Center would later be built. But our goal this day was to find out if it were even a possibility of him selling."

Mr. Cook refused them outright. In his opinion, $850,000 was not enough. To even consider the offer, he required $100,000 more. He would not only be selling prime property, he said, but giving up the rights to development.

"Jean and I were just as stubborn as Mr. Cook," says Lu, smiling. "We refused his demands." They knew, though, they needed to find a way to raise money, to try to convince Mr. Cook they were serious. Fund drives were fine, but they would never generate enough money to purchase the property. Lu returned to the county commissioners. She proposed that, if the Friends of Tryon Creek Park could somehow raise 25 percent of the funds, would the county consider putting in 25 percent, with the remaining 50 percent coming from federal sources?

The county turned down her down flat.

Undeterred, Lu came up with a long shot that nevertheless seemed plausible: state highway monies might be a way to fund 25 percent of the costs. To plead her case, she made an appointment to meet Glenn Jackson, chairman of the Oregon State Highway Commission. Jackson listened carefully and agreed in concept with the idea that the canyon would indeed make a splendid park. He was qualified in his enthusiasm, however. He said that using highway dollars to achieve the goal would only open up a can of worms all over the state. Disappointed, Lu couldn't contain her passionate response.

"I told him, 'But Mr. Jackson, in less than a month, Mr. Cook is going in front of the county commissioners to get his development plan approved!'"

Mr. Jackson replied cryptically, "Stay the course you're on." He added something that puzzled Lu even more. "He told me, 'If it looks like you are going to lose the battle, we will back you up.'"

Lu shakes her head. "I hadn't a clue what that meant."

Two weeks before the commission hearing, Lu received a phone call from someone she did not know. He said he was with the highway department in Salem. "Are you ready to go?" he asked her. "Are you sure you want to create a state park?"

Lu was flabbergasted. "'Of course, I do!' I exclaimed. But I had no idea what that man was talking about!" He told her he expected approval of federal funds that day. By 1:00 pm, he called to say funds had been approved.

Taking all sides by surprise, especially the county commissioners and Mr. Cook, Governor Tom McCall held a press conference later that day. Speaking to reporters, the governor announced that there was going to be a new state park in Oregon covering an entire 659-acre canyon. It would be called Tryon Creek State Park.

The following morning, the news hit the front page of the papers. It was sensational. Lu was aware that the battle wasn't won, though, whatever the governor said. The county's decision regarding Mr. Cook's proposal would be the determining factor. The hearing was set for two weeks hence. And the commissioners had made clear they believed constructing condominiums was the land's highest and best use.

The evening of the meeting, the county commission board room was packed. Mr. Cook was there, flanked by other developers all dressed in dark suits. But the overwhelming majority of the audience were advocates for a Tryon Creek State Park. Scores of supporters from the Kitchen Counter

Drive showed up. Students, teachers, neighbors, people of every age and from all walks of life had come to lend their backing, taking a united stand to "save a whole park."

The energy in the room was palpable. As Lu remembers, the planning director made an initial declaration. He relayed that the commission had received more letters in favor of creating a park than it had ever received on any other issue in its history. Mr. Cook, seated in the front row, was trembling with fury. The director continued, saying he had received a letter from Governor McCall himself, informing him that the state was going to put up the necessary funds.

Lu will always remember the director's next words.

"For these reasons," he pronounced, "I do not see how we can approve development on land which has been approved for a park."

"The fate of our canyon was sealed at last! The audience erupted in applause," says Lu, joyously. "That night, Tryon Creek State Park was born."

Thanks to an outpouring of public effort, the Tryon Creek Natural Area became the first state park in the city of Portland. Moreover, the Friends of Tryon Creek held the distinction of being the first citizen group ever to work to establish a state park in Oregon. As Lu says with a smile, the Friends' efforts didn't stop there. In 1973, over three hundred volunteers, under botanist Jean's supervision, began building trails in the canyon. They also helped raise money to construct a Nature Center in place of condominiums.

To Lu's lasting joy, Tryon Creek State Park became officially dedicated on July 1, 1975. In 1977, Lu became the first woman ever asked to sit on the State Parks Advisory Committee, later renamed the State Parks Commission, a position she held for fourteen years.

Today, at ninety-one years old, Lu still enjoys visiting the park, and feels the same zeal she has known since the first time she saw it. She says she is often asked, how did she do it? How did she stay the course, not get discouraged and quit outright?

"That answer is easy," says Lu. "The project was too important to give up. I knew the canyon and I loved it. With the exception of the developers and county commissioners, the community was totally supportive. That is what gives you the courage to go on."

Lu sits back in her chair. She smiles at the memories of events stretching back fifty years and her great pleasure in their outcome. "Even when things got hard, I could always picture the whole thing. Through it all, I could visualize the pleasure it would be for everybody."

If we want healthy farm land, we need to have
healthy habitats adjacent to it. If we want to
farm for the long haul, we need to maintain
a healthy ecosystem. We can't just manage
everything for the short term, because if
our practices over the next several decades
sterilize our soil, or we've mined our capital
out of the environment so that in ten, twenty,
or fifty years down the road it's all used up,
what is the point? . . . If we want to live on this
planet indefinitely, having an understanding of
nature is imperative.

―――――――――

Woody Wolfe

5

A Confluence of Values
WOODY WOLFE

Throughout the nation, "For Sale" signs are cropping up alongside farms and ranches at an alarming rate. Many farmers believe they can't afford to continue in business and won't have the ability to pass their lands on to the next generation. One Oregon farmer, though, is determined to buck that trend. Researching tools to protect his Wallowa County lands, to keep them agriculturally based and protect native fish and wildlife, he has successfully preserved his farm and the way of life he treasures.

Now he is working to let others to know how to save family farms and ranches.

∾

"It's worth it to me if I can protect this land forever from subdivision and population pressures and it stays in agriculture," says Woody Wolfe decidedly. "This year marks 121 years that the farm has been in the family. My sons are seventh generation. While nothing lasts forever, our goal is to take care of it and make it last as long as it can."

Woody speaks with a sense of respect for the landscape and with the easy familiarity that arises from being a part of the place that has formed him. The ranch has been a part of him his entire life. "I was taken home from the hospital to the house that I'm living in now," he says, acknowledging that one can't get much closer than that. He began a life of farming when a child, and it wasn't always easy. At nine years old, his dad had him running a large, articulating tractor.

"It was hard for me and stressful during wheat harvesting time, but I did it. My father later told me, 'I was sure that you could.' And I really couldn't afford to hire anybody else." He chuckles. "My dad always said, 'You've got to have them begin working before they go through the phase of knowing it all,' and he was right, really. It gave me a solid base that started while I was still inquisitive and liked to soak it all in. And today I still enjoy the work and getting things done, and making them work well."

Woody, though, admits that he wasn't sure, at first, that he would stay on the farm. He majored in agriculture when he entered Oregon State University, mostly to please his father. But soon, everything he was learning grabbed him. He started to focus and began realizing how much Wallowa County and the farm meant to him. The last year of college he met his wife, Megan. Woody graduated with honors in agricultural business and a minor in agricultural economics in 2002. He and Megan married that August and moved home to the ranch, taking over the business from his father.

In the beginning, he rented the ranch. To make the farm more economically viable, he endeavored to purchase neighboring properties. This, however, presented many obstacles. Quickly he realized that continuing to farm through family generations is not as easy as it sounds.

As Woody relates, his two-thousand-acre northeastern Oregon farm, classified as a mid-sized operation, lies within the range of farms and ranches that are disappearing nationally at the fastest rate. While family farms that size are arguably large enough to have a cash flow, they are not big enough to secure competitive edges in fuel, fertilizer, and other necessities to operate a profitable business.

But the matter has even more serious consequences. "Even if one wants to farm today, it's close to economically impossible to start out with nothing and build a functioning business without having a high source of non-farm income," Woody says. "That is unfortunate, because what it leads to is affluent people purchasing land as a hobby and a tax write-off. This generates inconsistency in our food production. You also have corporate farming driving out the family farm. Corporate farms continue to get bigger and bigger. Their objective is not necessarily sustainability, but to maintain the least cost economic structure."

The possibility that worries Woody most of all is rapidly spreading across the West. "Today, it's become more profitable to sell off farm land for development—for that million dollar trophy home or subdivision. This is especially common when farms change hands between generations. You can see it happening everywhere, even here in Oregon, as you drive around and see the 'for sale' signs popping up more and more. It's true; I could sell my land for several home sites and get a significant amount of money. And if this was not Oregon but Idaho or Wyoming, and this place had been 'discovered'—which I hope it never is—people like me would have huge amounts to gain. After all, I've got eight miles of Lostine riverfront property. Think how much that would that be worth in two to five acre lots."

That is not the future Woody Wolfe wants for his land. And while to some it may sound remarkable, what he desires is just the opposite.

"A percentage of our culture is starting to put a value on something besides money. Don't misunderstand me; I like money, and people who say money doesn't matter are generally people that have it. But it's not the number one thing in my life. I just need enough of it to make my life work. Once you have the basic necessities of food, clothing, shelter, and a degree of comfort, beyond that there is no increase in happiness. I think most people would say, if you asked them if they would rather be rich or happy, they would want to be happy," says Woody. "I am attached to the scenic value of my land. I don't want to see a lot of new homes going up. I don't want this valley to look like a suburb. My goal is sustainability—to keep the land healthy and thriving for as long as I can."

Woody and his wife, Megan, are blazing a new trail to allow them to retain the farm in the family and to keep its soil, water, wildlife habitat, and productivity sustainable. Woody hopes others who love the land will follow. It involves thinking about and managing farm and ranchland in a new and innovative way, using tools to help maintain cash flow in order to avoid selling the farm, bit-by-bit.

Woody is familiar with that former scenario. In the beginning he was forced to sell some of the land he bought in order to keep the rest. And he didn't feel good about it.

In 2002, Woody purchased 755 additional acres, which included a fair amount of irrigated, tillable land. He borrowed $500,000 from the bank and had five years to pay off the loan. "And there was no slam dunk on where that money would come from," Woody testifies. He made the first payment from the sale of two hundred cows that came already bred and were part of the purchasing price of the neighboring farm. In the next thirty months, however, he had to come up with the remaining $325,000.

Money was tight. To afford the second payment, Woody sold three hundred acres of his property that were buildable. It was later subdivided into two lots where homes now sit.

"From an economic standpoint, selling off the land it was a positive thing, but I just didn't have a good feeling. What I really wanted was to find a way I could generate money to continue farming, *and* feel good about it."

Before the next loan payment was due, Woody began researching options. One above all the others intrigued him. He knew very little about it, though. It was called an agricultural conservation easement.

As Woody discovered, a conservation easement is a voluntary, legal agreement between a land owner and a qualified entity such as a nonprofit land trust. It permanently protects property by restricting it for agricultural production, or for open space, recreation, or wildlife habitat.

"I had no intrinsic knowledge of what they were, so it was a learning process for me. Frankly, most people don't have a clue what one is or how it functions."

Woody knew that the land he owned held significant ecological and cultural value. The Lostine River, a picturesque, snowmelt-dominated system, provided prime habitat for wildlife and for migrating Chinook salmon, steelhead, and bull trout. Part of the ranch included the confluence of the Lostine and Wallowa Rivers—a section, he discovered, had historically been a Nez Perce summer fishing village.

In 2004, Woody read about James Monteith, a Wallowa County resident who had recently founded something called Wallowa Land Trust. The organization, in part, dealt with identifying valuable rural properties and helping landowners maintain and preserve them. Woody liked the stated mission of Wallowa Land Trust: "To protect the rural nature of Wallowa County by working cooperatively with private landowners, governmental entities, Indian tribes and local communities." And the way the Trust accomplished its goal was also agreeable: "[It uses] economic incentives to help conserve the valley's natural, historic, cultural and agricultural resources, including forests, farmlands, ranchlands, grasslands, wetlands, waterways and open space for the benefit of present and future generations."

Woody and Megan contacted Monteith and he explained that agricultural conservation easements were legal agreements between a landowner and a land trust, like Wallowa Land Trust, that acted to protect and preserve natural and traditional values of a property in perpetuity. In other words, a farmer or rancher could donate an easement and get a charitable tax deduction, or sell the rights to develop and subdivide his property to the land trust, continue to farm, and use the money gained from the sale to pay down debt and invest in the farm's operation.

As Monteith told Woody, Wallowa Land Trust was committed to keeping working lands working. The idea appealed to Woody and his wife. It was true to their values. An agricultural conservation easement, they decided, could help protect the rural way of life they loved, allow them to farm and keep the ranch in their family, and leave a lasting legacy for future generations.

Joining forces with Wallowa Land Trust, they went to work. Woody and Megan's vision was to pay off the loan to the bank, improve soil health on the farm, and enhance conditions for native wildlife and fish on their property, all the while continuing in agricultural production.

"It makes perfect sense to manage the land for ecological purposes," Woody avers. "If we want healthy farm land, we need to have healthy habitats adjacent to it. There is, I believe, a growing mental evolution within agriculture . . . an understanding that, if we want to farm for the long haul, we need to maintain a healthy ecosystem. We can't just manage everything for the short term, because if our practices over the next several decades sterilize our soil, or we've mined our capital out of the environment so that in ten, twenty, or fifty years down the road it's all used up, what is the point?"

Acquiring agricultural conservation easements, however, took far longer than Woody first anticipated—seven years to secure the first one and six more for the second. Finding funding was a large part of the challenge. To do that meant making partnerships towards a shared goal.

Together, the Wolfes and Wallowa Land Trust worked with charitable foundations, the Nez Perce Tribe, and with the Natural Resources Conservation Service through its Agricultural Conservation Easement Program (ACEP), a service that helps provide financial and technical assistance to farmers and ranchers with the goal to conserve agricultural lands, wetlands, and grasslands and to keep lands in agriculture. Woody and Megan donated one quarter of the easement value themselves.

Their hard work, however, paid off. The Wolfes' agricultural conservation easement was the first on a working Wallowa County ranch. Today, 463 acres of their working farm is permanently protected from residential development and subdivision.

"Our goal as a family is to take care of this land, have it stay sustainable, and provide for a sustainable business. Now it is protected forever. Whether my kids end up with it, or their kids, or somebody else who I don't know, these values are going to stay attached to the property, within this written document. Knowing that," says Woody, "I can sleep easier at night."

The Wolfe's conservation easement includes 2.5 miles of the free-flowing Lostine River and the confluence of the Lostine and Wallowa Rivers. The agreement placed strict controls to protect the habitat adjacent to the rivers. Taking it a step further, the Wolfes are currently involved in improving habitat for native Chinook and steelhead populations, in partnership with the Oregon Water Resources Department, the Freshwater Trust, The Columbia Basin Water Transaction Program, and the Nez Perce Tribe. The Wolfes also have a management plan, under an agreement with the Natural Resources Conservation Service, to use no-till farming practices, which provides better water infiltration, causes less erosion and disturbance in the soil, and results in higher organic matter content in the soil.

The easement also has a cultural component. Separate easement portions allow the Nez Perce Tribe to operate a fish weir on the Lostine River in the historic section once used by native peoples. This area now provides for a significant tribal fishery; all spring Chinook salmon harvested by tribal fisherman from the Lostine come from this stretch of the river.

"Today, we are getting a generational turnover in farms and ranches," says Woody. "Personally, I would like to see that be a much higher percentage staying in families, with less of the land being sold to developers. Partnerships are essential to creating success."

When farmers and ranchers can work with local, state, and federal natural resource groups, land trusts, and tribal leaders, Woody explains, it can help keep family farms viable. "They have research and science behind what they do; it's important to partner with people that have the capital and the knowledge to get things done," he says.

Woody is encouraged to see people moving to the county and becoming a part of an environmentally-focused community. "Entities like Wallowa Land Trust, Wallowa Resources, and the Nez Perce Tribe are hiring a lot of talented people whose main focus is not money. I like seeing people around the state raising their own honey, their own eggs, having flowerbeds with vegetables in them, recycling more, and doing the things that make sense to protect our environment."

Woody is fully cognizant that many farmers and ranchers do not take advantage of the tools offered in these kinds of partnerships—a fact he finds unfortunate. He understands, though, the mental block.

"Ranchers and farmers are independent," he expresses knowingly, a sixth-generation farmer. "In rural, conservative, agricultural communities, the term 'environmentalist' can be a bad word. Plus, the whole idea of the agricultural conservation easement process is fairly new in the West. Couple that with the fact that a rancher might be sitting down at the table with somebody they think may have chained themselves to a redwood tree in the seventies. That doesn't sit well. And for some, they realize that their property value is not going to increase as much as it would if it were still subdividable."

His voice deepens. "I would not like to see this place become a Sun Valley, Idaho, or a Jackson Hole, Wyoming. The net value of my property would go through the roof, but then, where would I live? I would not want to live here anymore. I would have to move . . . and I don't know where I would move to. What we still have today is a place where people know where their food comes from. They have a physical connection with what they eat because they grew it and have an emotional connection because it is their blood, sweat and tears."

The determination to ensure well-managed working lands be kept in production continues to motivate Woody to action. He is not alone in his sense of urgency. With the average age of the nation's farmers and ranchers at sixty years old, a massive transfer of agricultural lands is slated to occur in the next twenty years. Studies predict many family farms are poised to be lost forever, but in Oregon, that oncoming tide may be held back, if Woody and others like him have their way.

In 2017, a landmark bill to create the Oregon Agricultural Heritage Program was passed with wide bipartisan support by the state legislature and signed into law by Governor Kate Brown. It formulates a collection of strategies to help today's farmers and ranchers pass their legacies to the next generation, including providing funding for more agricultural easements on working lands—keeping them in production while preventing development and fragmentation of the land. Key organizations that crafted the legislation came from a broad, and often seemingly divergent, suite of factions: The Oregon Cattlemen's Association, the Coalition of Oregon Land Trusts, the Oregon Watershed Enhancement Board, Sustainable Northwest, the Oregon Farm Bureau, and The Nature Conservancy . . . all working together to preserve Oregon's agricultural heritage and the fish and wildlife habitat the land supports.

With the passage of the bill, a new Oregon Agricultural Heritage Commission program was formed. Twelve members were appointed, one of whom is Woody Wolfe, who actively lobbied legislators to get the program established.

"What was funded was the creation of the entity and new rules for the state of Oregon to allocate money to people for conservation easements, conservation covenants, conservation management plans, and succession planning," Woody explains. "Now we are in the process of public comment and approving the finalized set. Next, people can start submitting applications for grants, with the goal of protecting more agricultural working lands."

In a polarized world, the program is a beacon of progress and hopeful example of Oregonians from all sides of a dilemma coming together to find solutions for a shared goal—sustaining and preserving Oregon resources for future generations.

Woody is the first to say that the end result of placing a conservation easement on his land has been totally worth it. It took years of effort but preserves the land and the way of life he loves.

"To me, Wallowa County means home. Today, we have the technology and the ability to use our natural resources in a responsible way. If we want to live on this planet indefinitely, I believe having an understanding of nature is imperative."

Sometimes people ask, why do Oregonians define "public good" as something that takes nature into account? I think it's because we live amidst and with nature. It is incorporated into our blood. It is here, through our every waking moment.

_____

Barbara Walker

6

The Open Gate

BARBARA WALKER

In 1903, landscape architects John Charles and Frederick Law Olmsted Jr. had a vision for Portland: a "40-Mile Loop" necklace of nature to surround the city by connecting a series of parks. A century later, with the help of a woman who loved both history and nature, the Olmsteds' farsighted idea has become reality, but with one major difference. The 40-Mile Loop has grown to 140 miles, becoming an outstanding network of paths crossing parks, rivers, lakes, buttes, and forests, and encircling an entire metropolitan region.

∞

If fifty years ago Barbara Walker had not been a good typist, Portland, Oregon, would not resemble the place it is today. But Barbara *could* type—"in those days when there was no such thing as erasable paper, so you couldn't have any mistakes," she recalls, laughing. So, when a neighbor asked for her help in transcribing the minutes for a young organization called The Forest Park Committee of Fifty, she kindly agreed to lend her a hand.

Typing those notes would expose her to things she had never thought about before, causing her to see natural areas in an entirely new light. Eventually, Barbara Walker became one of the most passionate advocates for parks, trails, and open spaces the City of Portland has ever known.

Born in 1935 and growing up in the west hills of Portland, Barbara found constant delight in the woods and natural beauty surrounding her. "This has always been home," she says affectionately. "My parents, aunts, cousins and grandparents all lived here. When I was a little girl, I always walked to my grandparent's house who lived a few blocks away."

After attending local schools for her first seventeen years, she went to Smith College in Northampton, Massachusetts, graduating in 1957 with a degree in government and foreign relations. "I just loved it!" Barbara says, her face lighting up. "Many of my friends who graduated with me would go on to work at *The Atlantic*, *The New York Times*, and the *New Yorker.*"

Barbara, her roots always in Oregon, returned home after college. For a time she worked in an attorney's office, "where I learned to type" she says. Later, she took a job with the *Oregonian* and then the *Oregon Journal*, where she had her own column, Meanderings. "I loved writing the column," Barbara relates. "I loved it because I loved Portland history and Oregon history."

When Barbara married Wendell Walker in 1961, he was just starting a business of his own, and she switched her focus from writing for the papers to working for him. Before long, they had three children. Barbara thrived on the new adventure of raising three boys. When the children were still young, a neighbor, Dorothea Abbott, dropped by one sunny afternoon to ask for Barbara's help. Dorothea was the acting secretary of the newly formed Forest Park Committee of Fifty.

"Dorothea utterly hated typing," Barbara explains with a chuckle, "and she was at her wit's end. She asked me if I would take her handwritten notes and transcribe them for her. I said I would. I had no idea, then, what that would set into motion. It was from typing them that I became aware of the history of Forest Park. Dorothea had been instrumental in preserving the park by virtue of being one of its staunch early supporters."

As Barbara discovered, the Forest Park Committee of Fifty was a volunteer advocacy group representing different organizations—such as the Portland Garden Club, the Trails Club of Oregon, and the Western Federation of Outdoor Clubs—that had successfully worked together to create the park. After decades of striving for a goal that many said was impossible, Forest Park finally became a reality in 1948, preserving forty-two-hundred acres and the nation's largest urban wilderness forest. Seventy years later, the Portland city sanctuary has grown to over fifty-two-hundred acres.

Barbara was captivated by the story. As her understanding of Forest Park's value to the city grew, she became fascinated by the historic fight to save it. She learned that Forest Park had been originally envisioned at the turn of the twentieth century by prominent landscape architects, the

Olmsted brothers, whose father, Frederick Law Olmsted, had been responsible for the creation of Central Park in New York City. In 1903, John Charles Olmsted and Frederick Law Olmsted Jr. had been hired by the Portland Parks board to design a park planning study for Portland. At the time, the Olmsted brothers were well known and had completed numerous high-profile projects, including designs of numerous park systems, residential neighborhoods, universities, state capitols, and were doing pivotal work with the National Park Service.

Fueled by her love of history, Barbara was keen to learn more. With a good deal of sleuthing and perseverance, she at last ferreted out a copy of the original Olmsted report, published in 1903, long buried and forgotten. Looking it over, she recognized the document as remarkable and far-sighted. In their writings, Barbara discovered the Olmsted brothers had advocated for the creation of a Forest Park with a beautifully descriptive recommendation.

"A visit to these woods," they wrote, "would afford more pleasure and satisfaction than a visit to any other sort of park. No use to which this tract of land could be put would begin to be as sensible or as profitable to the city as that of making it a public park. . . . If these woods are preserved, they will surely come to be regarded as marvelously beautiful."

While the Olmsteds' plan was initially met with excitement citywide, not much was accomplished. Bond measures to carry out the Olmsted ideas repeatedly met with failure. Barbara was aghast at the revelations she uncovered. In 1912, changes were made to Portland's city government. Under a new charter, the Portland Parks board was no longer an independent entity; rather, park governance was to fall under an elected mayor and commissioners. The acting mayor at the time, Joseph Simon, strongly believed that creating parks was superfluous and a waste of money. The sad result was that the Olmsted plan grew dormant through the years and headed toward oblivion.

The greatest feat, as Barbara saw, had been accomplished by a handful of dedicated citizens who never gave up and fought for forty-four years to realize the Olmsteds' original vision for Forest Park—and won.

Barbara was astonished by the whole story. Grateful for the creation of such a magnificent park, she became curious about the other recommendations that the Olmsteds had offered. She continued her investigations. What jumped out at her was something that she had never heard about before. The Olmsted plan was far bigger than just Forest Park.

The great vision of John Charles and Frederick Law Olmsted Jr. was that the Parks of Portland *should be connected.*

Barbara remembers the first time she contemplated their rich ideas. "I realized the Olmsted plan proposed not only Forest Park, but recommended many places for parks . . . places along Marine Drive and the Willamette River, east to Rocky Butte, then extending to Washington Park, which was then called City Park, to Forest Park, to Council Crest and then all the way down to Terwilliger Boulevard. They envisioned them all connected! . . . Their thoughts were just so forward for that time. What was different was that they valued the land for what it was, not just what it could be used for. That is why their plan has had such a lasting effect."

As she studied the report, something else grabbed her attention. The beloved woods she had grown up with—Marquam Canyon in Southwest Portland—were also included in their plan, viewed as being worthy of preservation and becoming a park.

The timing of her discovery could not have been more fortuitous. Around the same period Barbara was to bring to light the Olmsted report, another plan for the area was already steamrolling forward. In 1968, three dentists were proposing an enormous housing development that would destroy the natural Marquam Canyon forever.

"It was one of those first planned unit developments, and it was huge. They were envisioning building a sprawling apartment complex, and that was only the beginning. It was long, extensive, and would wipe out the entire ravine."

Contemplating the destruction of the beautiful Douglas firs, western hemlock, red cedar, and big-leaf maples that clothed the wondrous woods surrounding Marquam Hill, and the scores of wildflowers that covered its floor in spring, was devastating to Barbara.

"It just seemed to me that the massive development shouldn't be there. Rather, it seemed so sensible to follow the Olmsted plan—to connect Council Crest and Terwilliger to Forest Park. The land was there; we just had to find a way to do it," Barbara says, her resolve still shining brightly.

"After typing up all those minutes of Dorothea's, I saw how a small citizen band never gave up the fight to preserve Forest Park. Now, it was our turn."

Barbara sought out neighbors she knew who loved Marquam Canyon as fiercely as she did. Soon, six friends sat around her kitchen table discussing how they might begin a campaign to save the wooded gulch. Included was Dorothy Westrack, a fourth-grade teacher at Ainsworth School; Nadia Munk, who had fled from her homeland of Czechoslovakia when the Nazis came to power; Elizabeth Crookham, an active civic volunteer; and Bill and Sonia Connor, he a doctor and she a nutritionist.

Each knew it would be a Herculean task. They decided their initial step was to form a small foundation. Barbara didn't know the first thing about foundations, but they gave it a name anyway—Friends of Marquam Nature Park. Next, the troupe began researching plat maps for tax lots numbers. Barbara dove into studying the geology of the ravine, which had exceedingly fragile soils, prone to slides. The group held meetings in each other's homes, discussing strategies to enlist other supporters. All agreed that the only way to attack the problem was to preserve the canyon, and the only way to do that was to raise the money to buy it.

The question of course, Barbara relates, "was how in the heavens do we do that?" For inspiration, the band looked to the success of Tryon Creek State Park, which had recently been saved from an enormous development by voluminous park supporters, led by Lu Beck. "If they can do it, so can we, we told ourselves," says Barbara. The Tryon supporters sold buttons and "feet of trail." Barbara thought that sounded like a good place to start. "So we made and sold little buttons that said, 'I'm for Marquam Nature Park.' We got lots of small donations."

The fledgling group compiled informational packets about Marquam Canyon, citing it was an oasis in the city. The packets described the critical mission to save the precious resource, which held scenic forested views and offered retreat from busy urban life. Quickly they discovered, however, that *distributing* the packets was what took the most time and energy.

"Remember, this was before the day of the computer or emails. Everything had to be done by hand, by mail, by copying machine, and word of mouth." And, in Barbara's case, by foot.

Barbara had learned that a major factor in Tyron Creek's preservation had been the support of Glenn Jackson, the chairman of the Oregon State Highway Commission. Jackson had helped elevate the issue to a state level, which eventually aided it in becoming a state park. While not acquainted with Jackson personally, Barbara knew from others that he was a tireless behind-the-scenes worker who supported parks and whose office was in downtown Portland. He was purported to be a businessman who didn't waste words and stayed long hours in his office.

Barbara's goal was to give Glenn Jackson a packet. She wanted to garner his interest, which she hoped might spark his endorsement of a Marquam Nature Park. She knew her timing must be impeccable to incur a minute in his busy schedule.

Using this reasoning, Barbara chose a day when Portland was enveloped by a major winter ice and snowstorm to approach Mr. Jackson. Gambling that he would be in his office, she donned her winter boots, stuffed the packet in her backpack, and trudged downtown. She didn't bother calling first, reckoning it wouldn't have mattered anyway. Phone lines were down because of the storm.

Mr. Jackson was indeed in his office when she arrived. When she opened the door, he seemed surprised to see someone but welcomed her in.

"I knew that he wouldn't be busy with someone else because no one could come to see him unless they had gotten there like I had—by hiking in the snow!" Barbara recounts.

Mr. Jackson listened. It was the first time he had ever heard of Marquam Canyon or of the idea of creating Marquam Nature Park. Barbara explained that the idea came from not just her or her small group, but from the original Olmsted plan to connect the parks. Even having a few buildings in the canyon, she entreated, would fragment it irreversibly. The Olmsted plan advocating for setting aside Marquam Canyon as a park seemed too phenomenal an idea to be lost.

Glenn Jackson liked the notion, but Barbara, not wanting to press her luck, asked him for only one thing at the conclusion of their brief conversation. Could he give her any names of people that he knew who might be interested in the proposal? Mr. Jackson, she says, graciously complied. With this list in hand, Barbara set out determinedly on her next mission: going door to door to meet them.

"When I would knock on their doors, I would ask for only five minutes of their time, no more. I didn't want to appear presumptuous," she explains with a sheepish grin. "You know, I was just a little housewife dropping by."

That housewife started getting positive responses to the idea. The people she met gave her more names of people, even corporations, to contact. If they didn't hop on the bandwagon right away, Barbara remained unfazed. "I'll bet they talked about this nutty woman who thinks she can do it—create a park! But then, eventually, many of them became great supporters and backers."

Contributions began to trickle in, then to grow. Donations came from neighbors, foundations, corporations, and businesses, along with matching grants. The Friends of Marquam Nature Park attended meetings with city commissioners, the planning commission, the geological commission and endured endless hearings, always stating their unified opposition to the three dentists' plan.

In time, a broad upwelling of community support began arising. More offerings came in. At last, the Friends of Marquam Nature Park secured the financial backing they needed to try to preserve the land for public use. Facing the overwhelming backing of creating a park instead of hundreds of apartments, the dentists at last relented, finally selling their interest in the canyon.

Marquam Nature Park, Barbara proudly beams, became a reality in 1978. In 1983, the park was officially dedicated. Six years later, in 1989, the Friends of Marquam Park transferred the two hundred acres to Portland Parks and Recreation. The Friends remain its devoted stewards, watching over the park to make sure it remains, as Barbara always envisioned, "an area of retreat from Portland's urban environment and an opportunity to walk in a natural oasis within the city."

Barbara's success in helping save a place that meant so much to her emboldened her to continue with her even greater aim, resurrecting the vision of the Olmsted brothers. "We needed to work to preserve the rest of the areas they'd imagined," she says. "Their plan just spoke to me. It seemed a natural to try to complete the dream! We needed to bring back their notion of the 40-Mile Loop."

The Olmsteds were advocating for the creation of a circuit of parks, all joined, making one giant greenspace that encircled the borders of the city. It was similar in design to a beltline for traffic, such as in Washington, DC, only this beltline was for people to connect with nature.

Barbara explains further. The Olmsted brothers' idea of joining parks was like stringing together jewels on a necklace. They pointed out many gaps that existed between the beads. What needed filling were those spaces between the parks. Springwater Corridor, Rocky Butte and Johnson Creek to Troutdale, and routes down the Columbia River and the Columbia Slough all the way to Rivergate needed to be procured somehow, and connected. The Olmsted loop would follow the Willamette River along Forest Park, joining Washington Park, Marquam Park, and Terwilliger Boulevard. There needed to be a link from Peninsula Point to the Burlington Railroad and on to Swan Island.

The list went on. When Barbara added it up, she realized it wasn't a 40-mile loop at all. In reality, as the city of Portland had expanded, it had grown to be a ring of parks and pathways, a natural ribbon, stretching a phenomenal 140 miles.

That didn't matter to Barbara. The name would remain what the Olmsteds gave it: the 40-Mile Loop.

"John Olmsted loved the views from Terwilliger. He thought we should save those wonderful lakes north of Portland, many of which are now gone, that adjoined the Columbia River." Barbara explains that the lakes have disappeared due to industrial development, dredging, piping, and levees to prevent flooding that changed the area's hydrology. "He thought we would be crazy not to do something with it. We should preserve it!"

Barbara reads from a copy of the old report that lyrically conveys their philosophy:

> *A connected system of parks and parkways is manifestly far more complete and useful than a series of isolated parks. . . . Parks should be connected and approached by boulevards and parkways . . . located and improved to take advantage of the beautiful natural scenery. . . . Scenic reservations, parks . . . and connecting boulevards would . . . form an admirable park system for such an important city as Portland is bound to become.*

It was not a dream for the faint of heart, nor for someone who did not have the stamina to take on a big picture, piece by piece, without losing sight of a much larger goal. It would take a giant dose of optimism and goodwill. Barbara was blessed with those virtues. She knew that securing each section of the loop would take time and work, bringing, at turns, successes and failures. In her favor she had a solid working knowledge now of the effort necessary to get things preserved.

The race was on. Big box buildings were erupting across the borders of Portland as the city expanded. Barbara's concerns concentrated on the lands around Marine Drive from Kelley Point and the Columbia Slough to Burlington Railroad and Swan Island. All those wonderful lakes dotted throughout were disappearing one by one. At risk, too, was natural habitat from Johnson Creek to Troutdale and on to Beavercreek.

Barbara's grand plan was to try to save them and then to connect them. Expanding on the Olmsted brothers' theme, her vision was to make it possible for people from all neighborhoods to go out their front doors and connect to trails. These trails would, in turn, join up with other trails, eventually even linking with pedestrian routes that could take walkers all the way to Mt. Hood, and from there, up and down the Pacific Crest Trail.

For thirty-five years, Barbara worked tirelessly to achieve that dream. At seventy-eight years old, she continues her advocacy still.

"People need places for refreshment and solitude and serenity. Our parks are not for carnivals; they're for people to enjoy what is natural in Oregon. We don't need to be entertained. Having something like Forest Park reminds us what a treasure we have. And then, we want to keep it! It makes people recognize nature, and seeing it, understand why it is so important." She smiles and adds something that underlies her truest beliefs. "As Oregonians, we get our fulfillment from the place, not from what somebody has built on top of it."

Today, through Barbara's indefatigable efforts, along with countless others who have been deeply inspired by her lively dedication, many of the pearls on that green necklace of nature in Portland have been acquired and preserved. A century after being proposed by the Olmsted brothers, the 40-Mile Loop now stretches 140 miles, joining together more than thirty parks, two counties, and six cities. It has been designated an Oregon Recreation Trail, an honor awarded to a single trail or a combination of trails that stand out for their outstanding quality and exceptional beauty and that define the essence of what is "Oregon."

Barbara, continually grateful, nevertheless expresses a note of caution. Breaches in the loop remain, she points out. An advocacy group, the 40-Mile Loop Land Trust, continues to pursue filling in the spaces. Incorporated in 1981, the nonprofit land acquisition organization, managed by a volunteer citizen board of directors, works with thirteen local jurisdictions to coordinate land purchases or donations, to accept conservation easements, and to act as a "land bank" of properties for future transfer to public agencies. They continue to meet with success in promoting the system of connected recreational rails, and strive to complete the Olmsted-inspired trail around Portland.

Barbara appreciates all the citizen efforts, frankly acknowledging that working to make these things happen isn't all fun. It can be tiring and full of disappointments. For her, though, she will never give up trying to close the gaps . . . and to open the gates.

"I can't think of how much time I have had to spend at hearings," says Barbara, shaking her head. "There are, and will always be, endless planning commission meetings. When you propose putting in a trail, people sometimes become apprehensive that it will bring the riffraff. They put up a fence along the trail. It doesn't take long, though, for people see how positive it is. Then, where there has been a fence someone has built, landowners put up a gate that can be opened to get to the trail."

Her eyes are bright. "Nothing is going to be more valuable than having a connection with nature in our cities, and an attachment to the earth. If we can keep nature in our cities, we can create the most wonderful communities ever."

Barbara Walker's amicable yet determined commitment has helped preserve hundreds of acres of natural beauty in the city she loves, making it accessible to all. Inspired by an old, overlooked report written by two men she never knew, she has helped make Portland one of the most livable cities in the world.

"There is motivation to extend what you love, and what you appreciate," she says thoughtfully. "Sometimes people ask, why do Oregonians define

'public good' as something that takes nature into account? I think it's because we live amidst and with nature. It is incorporated into our blood. It is
here, through our every waking moment."

Barbara's warmth radiates throughout the room. "Yes, I think my favorite idea of success is when someone cuts out a hole in the fence and puts in
the gate. To me, that is the spirit of Oregon."

You can't look at Oregon without getting
a sense of its primary values. There is something
in Oregon at its core that makes us want to be
protective. Oregon has priceless assets. You want
to say, "Don't blow it!" . . . There will be times
you are up against huge odds. This is when you
need to dig deep, and have internal fortitude and
commitment to something you believe in. And to
not let go.

———————————————

William Hutchison

7

Why Not?

WILLIAM HUTCHISON

The federal Wild and Scenic Rivers Act, signed into law in 1968, has protected fifty-nine rivers throughout Oregon, more than in any other state, keeping them dam-free in perpetuity. It would take a surprising decision by the Supreme Court of Oregon, however, in a highly unusual court case brought by citizens, a doctor, and a courageous attorney, to go one critical step further in upholding these rivers' integrity.

∾

A lifelong attachment to nature, for Bill Hutchison, began with something simple: the sound of water. But not just any water. While attending the University of Oregon, a friend invited him to go fly-fishing on the Metolius River, a glorious stream winding through a green valley on the east side of the Cascade Mountains. Bill was captivated. He quickly discovered that it mattered little to him if he caught a fish or not—just hearing the sound of the rippling river, seeing the colorful birds above, and watching the fish below transported his imagination as never before. The cold, clear water of the Metolius filled him with a love of place that would direct his future in ways he could have never envisioned.

A native of Portland, Bill grew up in a family of five lively boys. Early on, he emulated his father, who was always actively involved in the community, giving time to efforts he believed in. Bill took the role to heart. After receiving his degree in economics in 1963 from the University of Oregon, he worked for the American Field Service program in New York City for three years and then entered Willamette University to study law. It was there that Bill learned about the newly forming Oregon Environmental Council, or OEC—a nonprofit organization dedicated to ensuring clean water and air for Oregon, reducing pollution, and encouraging farmers to incorporate sustainable practices.

Driven by his growing passion for "doing what was right" for the natural environment, Bill got involved. Soon he became president of the Oregon Environmental Council. After passing the bar in 1969 and beginning work as an associate in a law firm in Portland, his involvement with conservation-minded organizations continued to expand. He volunteered for the Northwest Environmental Defense Center—a vigorous group led by Northwest College of Law at Lewis and Clark College in Portland—and was elected its president. Like the Oregon Environmental Council, the association is dedicated to protecting Oregon's environment and natural resources.

Bill believed in the charges of the organizations. "When you see these opportunities, it's good to get involved and try to influence what the ultimate result is. Because the extent to which we use our natural resources must be sustainable in perpetuity," he says, with genuine belief. "We see this in forestry, we see it in water resources, we see it in fisheries. We need to have a public interest component, not to supplant the economic activity, but to modify it to 'not-so-fast, not-so-much.'" After all, you and I try to live within our means; we have budgets. Why can't we do this for the environment?"

That credo has underlain Bill's motivation throughout his career. It is a principle he has lived by, even to the point of having to change law firms when his partners proved unsupportive of his volunteer activity.

In 1987, happily married, raising three daughters, and busy at work, Bill was appointed by Governor Neil Goldschmidt to chair the Environmental Quality Commission (EQC), a volunteer committee that oversees Oregon's Department of Environmental Quality (DEQ). The DEQ is responsible for issuing permits and penalties and adopting administrative rules statewide for the management of all of the environmental issues affecting clean water, clean air, and sewage treatment systems. For Bill, this was another way he could practice his public service. To allow him the flexibility to participate, he chose to work with a different law group that held values of volunteerism and environmental engagement similar to his own.

Bill's exposure to environmental issues gave him a personal perspective he continues to hold: the critical importance of sound science, and the belief that, in all decision-making whenever we are impacting the environment, there needs to be an integrated economic model—one that doesn't just look at a single parameter.

"Sometimes the 'sustainable in perpetuity' perspective seems to run head to head with economic development. But, looking more deeply, does it really?" he asks. "What is not always obvious is that many proposals and economic strategies only look at a single component: 'How many dollars did we make today?'"

Bill takes a wider view.

"We need to consider, 'What does this proposal cost in a holistic and true public interest sense?' This is crucial to consider, even when it is difficult to measure the invisible and difficult to quantify attributes."

This fact became evident in Bill's push for the City of Portland to construct its Big Pipe—the largest sewer assembly and public works project in Portland history, completed in 2011. Before then, stormwater runoff and raw sewage overflowed directly into the Willamette River whenever powerful rainstorms pummeled the city and overwhelmed the antiquated underground pipe system. To reduce the chance of serious overflows and increase capacity, a new, much larger system was proposed in 1991, after the city was sued for the discharges under the Clean Water Act. The plan called for building underground tunnels twenty-two feet in diameter and stretching for miles. East- and west-side tunnels would connect and take the sewage and runoff to the North Portland, Columbia Boulevard wastewater treatment plant. It was a costly proposition, with a price tag over one billion dollars.

The city, while aware of its sewer problems, let the project languish when initially proposed, citing it was too expensive and excessive. Bill disagreed and countered that thinking.

"Yes it's costly, but when you weigh it against the price of a river that won't work without it—so the fish can't live and you can't swim in it—you see that it's critically important to have decisions that reflect all the uses and all the values. There needs to always be holistic analysis of what the true cost of a proposal is, or what the consequences of failing to take protective measures really will be."

Bill advocated vehemently for the Big Pipe's construction. As his term as chair of the Environmental Quality Commission was coming to an end, the EQC ordered the city to upgrade and repair the city's system.

"And the city, to its credit, got up and built it and did it, on time and on budget," Bill continues. "In the long-term, the costs of inaction far outweigh the cost of action."

After its completion, combined overflows of raw sewage and storm water into the Willamette River were reduced annually by 95 percent.

The EQC faced numerous environmental issues while he was chairman. A second challenge concerned sewage problems, this time affecting the Tualatin River. In the late 1980s, Bill and the EQC stressed to officials the necessity of setting total maximum daily loads for the Tualatin. When speaking to the Washington County Commission and the Unified Sewage Agency, Bill made sure the commission understood why it was important to carefully manage the input of sewage into a river system like the Tualatin, which was slow-moving and therefore could not handle excessive daily loads. "You must do whatever it takes to operate within those limits," he remembers stating.

This directive, however, ran counter to the political and economic thinking at the time. The chair of the County Commission firmly repudiated Bill.

"That would be the end of development in Washington County," he declared.

"No. This will breathe life into Washington County," Bill replied. "If the river thrives, the economy also will thrive."

Bill and the EQC were right. And the effects of those decisions, as well as a million-dollar Clean Water Act lawsuit, changed the future of the Tualatin River. The Unified Sewage Agency, renamed Clean Water Services in 2001, built four wastewater treatment facilities, created stormwater management swales, and planted trees along the riparian zone. Today the water from the four treatment plants is the cleanest that goes into the Tualatin River.

In Bill's office hangs a quote by George Bernard Shaw that has always inspired him: "Some people see things and say why, but I dream things that never were and say, why not?"

"When someone says, 'It can't be done', well, I like to say, 'then we'll show you.' Of course, that is not always easy. There will be times you are up against huge odds. This is when you need to dig deep, and have internal fortitude and commitment to something you believe in. And to not let go."

Bill would never let go of Oregon. He believes one can't stand back and regard Oregon without developing an appreciation for its intrinsic worth.

"You can't look at Oregon without getting a sense of its primary values. There is something in Oregon at its core that makes us want to be protective. Oregon has priceless assets. You want to say, 'Don't blow it!'"

But we nearly did, when the City of Portland was poised to sell off pristine water from its Bull Run Watershed—the city's water supply located east of Gresham—for electricity, and sacrifice irreversibly the integrity of the Sandy River, a scenic river into which the Bull Run flows. Bill learned about the city's proposal when Dr. Arch Diack, a Portland cardiologist, approached him in 1986. Dr. Diack was deeply concerned. He and his brother Sam, an internist, had donated 160 acres of land along the Sandy River in 1971 to The Nature Conservancy to be used as its seed parcel creating the Sandy River Preserve.

The preserve is biologically diverse and has extraordinary qualities. The Sandy River is the last, undeveloped river close to a major city in Oregon. It is a designated Oregon scenic waterway and a national wild and scenic river. Part of the original donation is old-growth Douglas fir forest, with trees dating five hundred years old. It also exhibits an alpine influence, highly unusual and fragile in such a low elevation, but a result of the Sandy River Gorge's unique features.

Arch explained to Bill that the City of Portland was now proposing to divert water from the Bull Run River, a tributary of the Sandy, for electricity generation. If that were to happen, it would significantly impact the Sandy River's water quality.

Arch handed Bill a remarkable document written by Charlie Ciecko, the Multnomah County director of parks who had worked for years at Oxbow Park on the Sandy River. Ciecko had done a thorough analysis of the city's proposal and found records that revealed that what the city was propositioning was, by all accounts, wrong. Ciecko uncovered that the city's license for Bull Run water was only for municipal drinking water purposes—which is how it still stands today. Any part of the crystalline Bull Run that was not diverted for drinking water had been allowed to flow naturally into the Sandy River—which was designated a wild and scenic river.

The title of wild and scenic river is critically important. It is awarded to waterways that have been recognized for outstanding qualities—for remarkable fish and wildlife assets; cultural, historic, or geologic values; and natural beauty. When selected rivers receive this designation, they are to be protected in their free-flowing conditions.

The City of Portland, however, wanted to take advantage of the "extra" Bull Run water—that which was not being used for municipal purposes—and incorporate it for hydro power. By doing so, though, it had not considered the impact on the Sandy River, Arch told Bill. Or, if it had, it was ignored for the perceived economic benefit of the diversion.

Arch explained that the entire deal was going on behind closed doors, where plans were made between the mayor and the Portland Water Bureau. The plan they had in mind was to construct a reversible pump turbine to receive the Bull Run River, sell the energy to PGE—a regional power distribution company—and dump the water into an aquifer adjacent to the Columbia River in North Portland.

Arch was clearly alarmed, remembers Bill. "It is the Sandy River's best water. It's what really makes the scenic segment scenic," exhorted Arch. "The city is saying, they're just taking a little bit. Well, a little bit is a bologna slice. It all gets bologna-sliced away. You don't really notice it's going away, but pretty soon, there's nothing left."

As Bill attests, no one who ever knew Dr. Arch Diack could deny that he had an inspirational quality about him. He had also so thoroughly analyzed the problem, with Ciecko's help, that it seemed to Bill as if there was a real chance to do something. Bill agreed to represent Arch and some other conservation groups before the hearings officer for the Oregon Water Resources Commission. He also agreed to present Arch's unusual but creative visual to help underscore their position.

On the day of their appearance with the hearings officer, as Bill made his argument, he placed three mason jars on the officer's desk. Each was filled with water of different hues and clarity.

The first jar held clear, pure Bull Run water directly from the watershed.

The second jar held sparkling Sandy River water after the Bull Run River had entered and refreshed it.

The third one, as everyone could see, held dirty and murky water. It was Sandy River water, but before the cleansing influence of the Bull Run contribution.

In graphic fashion, the three jars represented just how critical the Bull Run River's influence was to the wild Sandy River. The Bull Run's contribution to the Sandy River made the beautiful and scenic waterway what it was. If its flows were substantially reduced, the Sandy would be completely degraded.

Nevertheless, the hearings officer rejected Bill and Arch Diack's argument and affirmed the city's permit, repeating a refrain Bill had heard many times before. "It's taking just a little bit." The officer concluded that there was nothing that said the city could not divert water, also citing that the part of the Bull Run River that the city wished to use was above and outside the designated scenic waterway.

Arch, however, was not one to give up. Nor was Bill. To Arch Diack, the Sandy River was his cathedral.

"Hell no, over my dead body they're going to do that to my river! We are going to protect the Sandy's scenic qualities," Arch proclaimed.

Bill appealed the hearings officer's decision. They took the case to the Court of Appeals. They lost there, too. The Court of Appeals applied the same reasoning as before: the city was only requesting to take a "relatively small portion of the total" flows, and within its rights to take the water.

Eventually, in 1988, Bill and Arch took their case to the Oregon Supreme Court. Their foundational argument stayed the same: had the City of Portland, with the water bureau's authorization, gone beyond the scope of its permitted license, and what would the effect of the city's diversion of water, taken above the scenic waterway, be on the protected stretch? The Supreme Court was their final chance.

While the basis of their case remained unchanged, Bill recounts, what happened at the Supreme Court changed history. Arch had come up with a stroke of genius. He passed his idea along to Bill. Bill agreed to give it a try.

In front of all seven justices of the Oregon Supreme Court, Bill asked to use a prop. Not the three mason jars, this time. Rather, as Arch realized, sometimes a picture is worth a thousand words.

In the midst of his oral argument to the court, Bill was granted permission to pick up a handout from his desk. Removing some sheets from an envelope, he turned back to hand a piece of paper to each Supreme Court justice. It was a reproduction of the Mona Lisa. The portrait, though, had one glaring difference.

Mona Lisa's lips had been cut out.

With that omission the entire masterpiece had been changed.

"We cut out "just a little bit," Bill explained to the court as they studied the picture. "Of course, it is an essential part. It's just like taking out Bull Run, the Sandy River's best water."

In the end, the Oregon Supreme Court ruled unanimously in Arch Diack's favor. In a historic action, on July 26, 1988, the court reversed and remanded the decision, effectively denying the appropriation of Bull Run water for electricity due to the effects on the Sandy River waterway. It ruled that the Bull Run River's flow could not be reduced and thereby negatively impact the Sandy River, a downstream a scenic waterway.

The monumental judgement that was called and still remains the *Diack* decision held implications for all rivers in Oregon, not just the Sandy. It was foundational for any river with a scenic river segment designation. The Oregon Supreme Court held that, contrary to the Oregon Water Resources Department's position and interpretation, the Oregon Scenic Waterway Act—ORS 390.835—meant something when it said that the "legislature declares that the highest and best uses of the waters within scenic waterways are recreation, fish, and wildlife uses. The free-flowing character of these waters shall be maintained in quantities necessary for recreation, fish, and wildlife uses."

The court concluded that since "the Sandy River is a designated scenic waterway, this designation requires that the free-flowing character of the Sandy River shall be maintained."

What Bill and Arch and others had achieved, with the help of a slightly altered portrait, was a victory for all free-flowing rivers. "Diack flows" would become the official terminology for quantifying what a scenic segment needs. Before any out-of-stream appropriations above scenic river segments can be allowed, "Diack flows—the amount of water essential for fish, wildlife and recreation—must be determined, established, and take precedence. The ruling was a victory for the Sandy River and all scenic waterways in Oregon.

Bill counts this ruling as the highlight of his career, as does Multnomah County Parks director, Charlie Ciecko, who researched the penned the initial white paper that he gave to Arch. Bill cautions, however, that even with such laws enacted, the work is not over.

"One thing I have experienced is that you can't win something like that and just go home and read the paper or leave town. All around the state, particularly in Central Oregon when the department was developing its Diack flows, the resolve can be extinguished."

Opposing interests often look for ways regulations can be skirted. The *Diack* decision is no exception. Bill elucidates one way that has been attempted to try to reduce flow guidelines is to advocate for "average sufficient flows." This, however, ignores the fact that there are certain amounts of flows that are critical at different times of year.

"So we get a lot of water in the river when the fish don't need it, and none when they do?" asks Bill. "And that's okay? No, it's not."

Bill is clear that one needs to be ever-vigilant. Battles and court cases still ensue over the application of the *Diack* decision. That is why environmental advocacy groups, notably WaterWatch of Oregon, says Bill, are of such value. Even before the *Diack* decision was enacted, WaterWatch of Oregon has pursued a mission to protect and restore water to Oregon's rivers for fish, wildlife, and people. WaterWatch is one of the groups Bill points to as an organization that works hard to keep regulators accountable and to protect Oregon's key water resources. Acting as a watchdog for the public interest, WaterWatch endeavors to enforce laws designed to protect fish and to allow future generations of Oregonians the opportunity to see free-flowing rivers and lakes, not dry stream beds.

"I would be despondent about a model that resulted in Oregon becoming like other places," says Bill, "with overburdened ecosystem environments and degraded natural habitats that can't sustain us and provide the benefits that we value. In ten to thirty years I hope that Oregon will not have fallen victim to overpopulation and more development than its natural resources can accommodate.

"Oregon is driven by its sense of place. There's a fierce independent spirit here, such as has been manifested by former governor Tom McCall. It's a place where you can adopt a Bottle Bill, requiring a deposit for returnable cans and bottles; you can adopt urban growth boundaries to protect Oregon agriculture and confine urban sprawl; and there will be a lot of fighting about it, but the value of those decisions is so profound it proves Oregon's core values."

His tone becomes more serious. "Be careful, be conservative, and be cautious—particularly about how we treat the environment. If we can't get that right, it will be to our peril."

Bill pauses. He then returns to his cheerful demeanor and the "why not?" inner confidence that has always defined him. At seventy-five, that optimism is still a driving force, a fact confirmed by his wife Barbara after more than fifty-four years of marriage and by his three daughters who adore him.

"I'm not sure, but there's just something in me—perhaps a sense of intergenerational equity—that makes me want to do what's right, to try to make things better. What would I like to see for Oregon's future? I would love to see Oregon be able to win that balancing act that venerates our intrinsic values—the values we really want to sustain."

If it does, Oregon owes much to a man who devoted his life to protecting those ideals in his positive, good-natured way. "I've always believed," Bill says with a smile, "that the last place to be able to offer quality of life will win."

I have always believed if you have more than
you need in life, then you should share it.

_____

Maribeth Collins

8

Reaching Across the Aisle
MARIBETH COLLINS

A pileated woodpecker gives its rousing call from a tall Douglas fir tree nestled in the middle of the forest. A family of four stops to listen. The look of wonder on the small girl's face who hears it is exactly the reason Maribeth Collins gave land in northwest Portland to be set aside for the birds. It was a gift that would later become the impetus for the first foundation in the state dedicated to acquiring, preserving, and restoring Oregon's parks.

∽

At age ninety-two, Maribeth Collins sees what our country is sorely lacking in the twenty-first century. It is something the native Oregonian first learned as a child, growing up with a Methodist minister father who was a passionate crusader for temperance work. Reflecting on that time, she returns to a belief she has carried with her all her life.

"It is always better to try to work with people, than work against them. I learned that from my father and Clarence Darrow." She smiles. "They disagreed on practically everything."

Clarence Darrow, an attorney who practiced law from 1878–1938, is considered one of the greatest criminal defense lawyers in American history. An outspoken agnostic, he was a leading member of the American Civil Liberties Union.

Maribeth's father, Clarence Wilson, was a critic of many of the issues Darrow championed. In 1924, Clarence established a headquarters on Capitol Hill in Washington, DC, for the Board and Church Society of the United Methodist Church. The building, still there, is the only nongovernment building on Capitol Hill.

"My father championed Prohibition," relates Maribeth, "and stood on the opposite side of the prohibition question with Clarence Darrow. During the Prohibition era, from 1920–1933, he debated with Darrow all over the country." In the nationwide arguments, they would vehemently attack one another on the platform. But after the debates were over, Maribeth adds, "they would sit together for hours, discussing everything—because they liked each other so much. My father and Clarence Darrow formed a lasting friendship. They may have had many things they disagreed on, but they really respected one another. Today, we are all seeing a Congress where nobody can go across the aisle. In the past, I observed that, while there could be disagreement, people still were friends, reaching across the aisle." Maribeth pauses, then adds thoughtfully, "I think our country needs so much more of that now."

The overarching philosophy that Maribeth observed during her childhood has remained with her. She freely admits it has been a driving force behind all her choices, her accomplishments, and the legacy she and her husband, Truman W. Collins Sr., have given their four children: to try to work together to make where you live a better place for all. As time would show, her beliefs would have a fortunate beneficiary: Oregon.

After spending her early years in Washington, DC, Maribeth returned with her family to Oregon when her father retired in 1936. It was a happy homecoming to the place where her pioneering mother's family settled in 1852. "And I've been here all the rest of my life," she says proudly. In 1939 she attended the University of Oregon, majored in literature, and upon graduation, took a job where she could be surrounded by something she loved—books—at a small neighborhood library in Portland. It was there that she first met her future mother-in-law.

"Truman's mother came in one day to check out some books. When she came back, several days later, she brought her son with her! She introduced us. Before long, Truman came in again, and then again, to check out and return books. That's how it all started," Maribeth shares with a smile.

What attracted her to Truman Collins right away were the values he professed. She learned Truman was a third-generation lumberman who had family ties in Oregon dating back to 1855, the same time her family had come to Oregon. The more they talked, the more she saw that they shared similar ideals about what was important in life.

"Truman had a dream that he wanted to establish a permanent community with sustainable forest management. He was really committed to this idea, long before anyone really had heard about it," she explains.

"After graduating from college and finishing graduate school in business at Harvard, he managed two mills in the traditional way. Eventually, they began running out of timber. Truman observed other places—small-town mills that logged timber all around them—and saw they, too, would run out of trees, close down, and move on to the next town."

This deeply troubled him, Maribeth relates. "Truman became aware what happened to the towns and families when their mills shut down. It uprooted everyone. It was difficult on all the people who lived there. He was determined to find a model that didn't do that, one that was managed so it could go on and on and on."

Truman told Maribeth about a professor he'd had at Harvard who had talked about something called "sustainability." The more he began learning about a new system for forest management that was different from the status quo, the more it inspired him to action.

"Truman was engaged in building a mill in Chester, California. He wanted to follow that model. He explained that it had to be the right size, and work on a truly sustainable basis. They would never, he affirmed, cut more than they could grow. They would work to keep a natural forest."

Truman's ethics were a match to her own principles. Before long, the gentleman sixteen years her senior won her over. They married on March 12, 1943, when Maribeth was twenty-four years old.

"Truman was ahead of his time. He loved Chester. He loved the people. He started a scholarship fund for the community. Today, the mill is still operating sustainably and so is the scholarship fund," she says with pleasure.

Maribeth considers how she first became interested in such matters as forestry. "Truman led me to understand the need for sustainability, and what lay behind his motivation. I soon took on that vision, and together we became dedicated to carrying it in all that we did, and all we still try to do."

Such forward thinking led the Collins Companies to achieve national recognition for their forestry management. In the late 1990s, they became the first privately owned forest products business in the United States to have all of their hardwood and softwood forests certified by the Forest Stewardship Council—an organization that promotes environmentally appropriate and socially beneficial management of the world's forests.

Maribeth and Truman both strongly felt they wanted to give back to the community they cared about in a long-term way. To meet that end, in 1947, when Maribeth was twenty-nine years old, Truman had documents drawn to create The Collins Foundation. One of the first foundations in Oregon, its goal was simple yet profound: to improve and enrich the state of Oregon, the quality of its natural resources, and the life of its people.

For the next seventeen years, while Truman diligently worked to build the company and foundation, Maribeth remained busy supporting his efforts and raising their three young children. Then, after twenty-one years of marriage, when she was six months pregnant with their fourth child, Truman Collins Jr., tragedy struck. Without warning, Truman Sr. died of a heart attack. It was 1964, and Maribeth was forty-five years old.

In that instant, Maribeth's life changed forever. The decisions facing the young widow with four children, one a newborn, were immense. The future of the Collins Company was suddenly in Maribeth's hands. She could sell the company and end the family's legacy after three generations. Choosing that option, she could concentrate on herself, seek the solace of privacy, and let that part of her life go. Alternatively, she could carry on the vision of her husband, Truman Sr., and their shared dreams.

The choice was not difficult for Maribeth. She would never sell. Instead, without knowing exactly what it might mean, she stepped up to become the chairman of the board.

"Then something happened next, to my utter amazement," Maribeth offers. "Truman's brother and sister, who were trustees of the foundation he had formed, came to me. They said they wished me to be the president of the Collins Foundation. I told them, 'Why, I've never been to a meeting!'

They assured me they would help me all they could." Maribeth's first reaction was "I can't do this!" Then she searched deeper in her soul.

"I realized, if I could learn, I would be carrying on something that meant a lot to my husband." Maribeth's erect posture reveals her determination and spunk, still evident after nine decades. "So, I plunged in!"

In 1964, she took the helm of the Collins Foundation. At the time, it was smaller and the work less formal, she explains, and primarily required that she go to her brother-in-law's office and read grant requests. When there were applications enough for a meeting, she called for one. Through the years, the workload increased and so did the requests. Maribeth continued to rise to the challenge, finding the work especially gratifying.

"I have always believed," says Maribeth, "if you have more than you need in life, then you should share it."

Grants from the Collins Foundation have generously supported Oregon arts, education, health and science, religion, humanities, youth, and the environment. They also resulted in the creation of another new foundation, the first of its kind in Oregon: the Oregon Parks Foundation. In the early 1970s, Maribeth donated an eight-six acre piece of property to the City of Portland to be used as a bird refuge. Located five minutes from downtown, the natural area lay in the heart of Balch Creek Canyon, a picturesque natural drainage of ecological importance adjacent to Portland's Forest Park and the Audubon Society of Portland.

The generous gift landed on the desk of a young city attorney, Don Jeffery. At the same time, he was contacted by a new Portland Park employee, Bob Bothwell. In discussing the donation, they both realized that the gift could be the seed for something much bigger. It could be the beginning of a new vision for Portland parks, in the manner of a challenge grant to create an endowment to help acquire and preserve Portland natural areas. Even more, it could be the springboard of a foundation that could provide grants and scholarships to support environmental, recreational, and educational improvements to urban parks.

Receiving permission to go ahead with the idea, Don drew up legal documents. Both he and Bob felt it should be incorporated as a nonprofit. They called it the Metropolitan Parks Foundation. Governor Victor Atiyeh liked the idea of a parks foundation for the city and supported it. It passed the legislature, paving the way for the final step: the Collins Foundation placed its property into the new parks foundation as the anchor. Subsequently, the donation was used as matching fund money for federal grants, thereby doubling its reach.

In 1979, the Metropolitan Parks Foundation took a new step, still under the guidance of Don Jeffery. Recognizing the need throughout Oregon for support for parks, the nonprofit was expanded to include the entire state. Don drew up the new charter for a statewide foundation, which became known as the Oregon Parks Foundation. Maribeth Collins was a founding board member.

Today the Oregon Parks Foundation has gifted millions of dollars to support Oregon's parks and native landscapes. In 2009, it became a dedicated fund of the Oregon Community Foundation, still maintaining its purpose of park acquisition, preservation, and restoration. Owned by Metro and managed by the Audubon Society of Portland, the Collins Sanctuary—Maribeth's catalyst bestowment—is still "for the birds" and serves as an important preserve for native birds, to the delight of birdwatchers.

To Maribeth's gratification, the Collins Foundation has aided many of the places she cares about in Oregon. It has administered numerous grants to The Nature Conservancy for the protection and management of land in Klamath Falls critical for biodiversity, and to northeastern Oregon's Zumwalt Prairie, known for its abundance of wild buteo hawks, which is the last, large, native bunchgrass prairie in the nation. It is especially meaningful to Maribeth that her family's contributions can help benefit Oregon, the place she loves best.

"I would rather live here than anyplace else in the country," Maribeth declares. "My roots go deep in Oregon. I believe it is important to have natural areas that are preserved for people to enjoy. Oregon has wonderful and beautiful things—and people respond to that. I respond to that! I am committed to preserving natural areas. I feel tremendously fortunate that we have been able to help, in some way, to do that for the people of Oregon."

For Maribeth, her values began with her parents—especially seeing the father she loved and respected reaching out across the aisle. From there, they continued to deepen as she understood the importance of searching to find that middle ground. Working first with her husband and later with her family, she has observed, over and over again, when people come together to try to discover solutions, there can be results that can benefit everyone. The important thing? To embrace people, all people, unconditionally.

"Lately I have been thinking about my life. There is a quote that has always held meaning for me, and how I have hoped to live," she says. "It's by William James." As Maribeth speaks, she smiles. Her eyes are clear and her face is sweet.

"The great use of life is to spend it for something which outlasts it."

A spiritual elder once said, "When we have healthy fish, then we will have healthy deer and elk, healthy birds and plants, and a healthy climate for ranching, farming, and logging." With our global need for water, conserving our watersheds is crucial. The water, I believe, should go to the fish.

Joe McCormack

9

Three Trails

JOE McCORMACK

In 1999, only thirteen Chinook salmon were captured on a weir at the site of a historic Nez Perce fishing village on Wallowa County's Lostine River. Twenty years later, after dedicated efforts of Nez Perce tribal biologists working collaboratively with farmers, ranchers, and the Oregon Department of Fish and Wildlife, over two thousand Chinook salmon are now returning home each year.

∾

The mystery and allure of the snow-capped Wallowa Mountains far to the south of his childhood home always held a fascination for Joe McCormack.

"In the 1940s we lived in the West Spokane, Washington, plains," says Joe, a native Nez Perce. "My father always talked about the Wallowas. He'd tell us stories and say to us that's our home. I always wanted to come here and see it for myself."

It would be years later, though, and after his father had passed away, before Joe would travel to the place he had heard so much about. While the curiosity to visit the land of his ancestors never left him, Joe loved roaming and exploring the area around his home when he was a child. West Spokane was a rural and diverse community, says Joe, and with six brothers and sisters he was never lonely. He knew his neighbors, for they were always coming and going and visiting; his father was very outgoing and popular, says Joe, and loved people. In his youth, his dad had also been a tremendous athlete, which, Joe explains, is the reason his family lived away from the Lapwai Nez Perce reservation in central Idaho, where many of his cousins, aunts, and uncles ranched and farmed.

"My dad was a full-blooded Nez Perce, born on the Indian reservation in Idaho. He was a gifted ball player, and my grandparents shipped him off to school at Clarkston, hoping that he'd get an athletic scholarship to Washington State University. My father did, and played baseball there for three years. He went on to play Pacific Coast League Ball. After the war, he signed with the Spokane Indians, a Triple A minor league baseball team."

Then his father's life took a tragic turn.

"On the way to play a team in Tacoma, Washington, the bus they were travelling on was crossing the Cascade Mountains over Snoqualmie Pass when it was forced off the road and lost control. It rolled down a three-hundred-foot cliff. Nine ball players were killed in the accident."

Joe's father retired from baseball soon after that and returned to Spokane. He met Joe's mother and married. They moved to the western plains where Joe was born.

Remembering his childhood, Joe smiles. "We were close with all our neighbors. There was a Chinese family, an Italian family, a Swiss family, rich people and poor people, ranchers and farmers and businessmen. I was lucky to have been in all of their homes and to hear the praise they would say about my father. Above all, I remember them saying, "Your dad always had room for our feet under his table."

The plains—with wide-open spaces, farms, forests, and fields—reso-nated with Joe. "For as long as I can remember, I have been drawn to the countryside. When we'd visit Spokane, I even loved the city parks." Joe en-joyed learning about the birds and trees. His Swiss neighbors, the Flabians, who owned and managed a small forest, had large gardens and kept bees and taught him names of birds and plants. They explained to him how they managed their trees. From this exposure, Joe got an early perspective on resource management and developed a keen interest in studying nature.

The path leading to Joe's passion and career in natural resource protec-tion would take many more turns and three decades before developing into what would later become his life's purpose. In the meantime, though, some-thing happened that changed everything for him. When he was a freshman in high school, his parents moved away from the rural western plains he loved and into town.

"It broke my heart," says Joe. "I rebelled."

Joe says he did poorly in high school. After graduating, he joined the US Marine Corps and did a tour of duty in Vietnam. Returning to Spokane, he attempted college, "but with the Vietnam experience, I didn't do well in being able to focus."

Joe tried the trades, bouncing around until a close friend of his cousins offered him a job as a carpet layer. Joe apprenticed under him for four years and when his boss retired, was able to keep his accounts and business under his own name. Joe continued to work as a carpet layer in Spokane for nearly twenty years. Still, rural areas haunted him.

"There was a big blank space in my life; something was missing," Joe recounts. He wanted to learn more about his father's side of the family and his Nez Perce ancestry. When in his late thirties an opportunity arose to live with a cousin of his father's and help with tribal fishing on the Columbia River, Joe made the decision to quit his business and moved to Rufus, Oregon.

Joe enjoyed fishing. Even more, he found himself increasingly intrigued by his cousins' participation in an old Indian religion that he knew nothing about, Waashat. As Joe was to learn, Waashat was basically a way of life. Some people called it the Seven Drums or the Longhouse Religion. It included ceremonial music, dancing, worship, and ritual. Before the coming of white settlers, it had been commonly practiced throughout the Pacific Northwest.

Joe stayed on the Columbia River and by the late 1980s, started fishing on his own, then with his brother who had come to join him. When times were slow, Joe worked construction jobs during the day and in the evening acted as a bartender at Bob's Texas T-Bone and Frosty's Lounge in downtown Rufus. It was there while tending bar that Joe met two people who would transform everything about life as he knew it.

The men were professors from the College of Southern Idaho in Twin Falls. They were travelling through Oregon on their way to the University of Washington, where they were involved in research on raising sturgeon in captivity. It was a challenge, they told Joe, and thus far scientists had reaped little success.

Joe enjoyed discussing the topic with them. "They explained how scientists could never figure out how to breed sturgeon. Sturgeon spawn over a long period of time, unlike salmon who will spawn, lay eggs all at once, and then die. A sturgeon lays its eggs over a period of a month or more and in different places. So, in captivity it's hard to get a sturgeon who will lay eggs, and lay them in a time that's manageable."

The professors took a liking to Joe. They explained that the College of Southern Idaho was a community college that focused on agriculture, but had an aquaculture and fisheries component.. They told him he should consider going to school to become a fisheries biologist.

Joe, laughing, told them no.

They came back the following year. Meeting once more at Frosty's, the professors expressed the same thing—urging Joe to attend college. Joe declined again. They returned a third and fourth year, still hoping to light a fire in Joe to take classes. Wearing down a bit, Joe confessed to them that the timing of the college probably prohibited it. After all, he fished in Alaska beginning in mid-May and returned home to fish on the Columbia River until late September.

The professors didn't budge. "They told me, 'Come on down. We'll make it work.'"

Joe at last acceded. He left Rufus and went to the College of Southern Idaho, majoring in fisheries biology while also taking classes in math, English, and writing. He smiles. "I graduated when I was forty years old."

Soon after graduation, in 1995 Joe was offered a job with the Nez Perce Tribal Fisheries in Wallowa County. Tribal Fisheries worked with the Oregon Department of Fish and Wildlife and were partners in the management of the county's fish resources. Joe was swift to take the position. He keenly remembers the day and the emotions he felt upon arriving in Wallowa County, the place of his ancestral homeland, seeing the snow-capped mountains encircling the green valley quilted with fields of alfalfa. The scene aroused in him a sense of awe and wonder.

It was a feeling Joe would later describe as *Tamkaliks*.

"Tamkaliks is a Nez Perce word that, in simplest terms, means 'from where I can see the mountains.' It is much more than that, though," Joe reflects. "Our tribal linguist has said Tamkaliks is more than a word. It is the feeling you get when you've been told where this place is, and you come over that horizon. You see the mountains and that valley. It is the emotion that makes you stand up and take notice."

Joe did take notice, and the new awareness awoke something inside him. His mindfulness did not only go in one direction, though. Beginning work as a tribal resource specialist, others took notice of him.

At the time, only two or three Nez Perce lived in the county. When people discovered that Joe was a descendant of Chief Joseph's band, several locals contacted him. "Right away I became involved with something that was just getting started, which would be later be known as The Nez Perce Wallowa Homeland Project," he explains.

The Homeland Project was an idea conceived by a Nez Perce named Taz Connor, who lived in the county and worked for the US Forest Service. Taz was a decorated Vietnam War combat veteran and a direct descendent of Old Chief Joseph. The dream he held was to restore, in some way, part of Nez Perce culture in their native homeland. Two years before, in 1993, Connor had organized the first Native American festival in Wallowa County. Called the Wallowa Band Descendants Powwow and Friendship Feast, the celebration was held in Wallowa's high school gymnasium.

Part of the vision, as Joe learned, was to create a place where the Nez Perce could honor ancient traditions and customs in the land that they once called home. They aspired to build a center where different cultures could come together to foster understanding. Beyond that, the idea provided a meaningful way to bring healing to a painful part of United States' history, when the Nez Perce were aggressively forced from their homeland, where they had lived for over nine thousand years, and into exile.

Joe was well aware of the story; his ancestors had lived through it. Briefly he explained the sorrowful time. For centuries the Nez Perce had lived on 17 million acres of lands on what later would become Oregon, Idaho, and Washington. The Wallowa Valley in Oregon, where Joe's family had come from, was an essential part of the Nez Perce homeland.

Nez Perce had first encountered the white man when the Corps of Discovery made its expedition in 1805 and 1806. Captains Meriwether Lewis and William Clark would repeatedly write in their journals that the Nez Perce offered them good will, aid, and friendship. Just seventy-two years later, however, the alliance turned into bloodshed and betrayal. As white settlers began arriving, in 1855 a treaty was signed by the chief of the Wallowa band, Old Chief Joseph, and the US government, allowing the Nez

Perce to stay in their valley. But more and more people continued to come. In 1863, the contract was re-negotiated by Nez Perce tribal leaders and the US federal government in Idaho.

Unfortunately, this document was created without the consent of the Wallowa band, and with devastating consequences. The pact ceded the entire Wallowa Valley to settlers and demanded the removal of the Nez Perce who had resided there for generations.

Old Chief Joseph was heartbroken. Before he died in 1871, he entreated his son to promise that he wouldn't allow the US government to take over their beloved lands. Young Chief Joseph vowed to honor his father's wishes. In 1873, Young Chief Joseph was able to re-negotiate with the US government, and they came to a mutual agreement. The government treaty declared the Nez Perce could stay on their land in the valley.

Four years later, however, without dialogue with the Nez Perce, the government abruptly reversed its policy. In a stark declaration in 1877, the US government demanded the Nez Perce Wallowa band leave the area. If they refused to abandon it, Army General Oliver Howard threatened to attack them, backed by his two thousand troops.

Joe explains that the Nez Perce have always been a peaceful and gentle people. Young Chief Joseph was no exception. Choosing peace over war, Young Joseph made the wrenching decision to leave their valley for the sake of his people's safety. But he refused to allow his people to be marshalled from their home and forced to an unknown place and future. In a desperate move, Young Chief Joseph led twenty-nine hundred Nez Perce men, women, and children on a path to escape to sanctuary in Canada, where they hoped to live in peace.

War erupted. The US Army pursued the Nez Perce on a disastrous journey of 1,170 miles. Only 40 miles from the safety of the Canadian border, the Nez Perce were captured. There were grievous losses. Over one thousand Nez Perce, including women and children, and three hundred soldiers lost their lives.

The remaining Wallowa Band was forced to a reservation in Oklahoma for ten years, where Young Chief Joseph continually advocated for their return to the Pacific Northwest. Finally, in 1885 the band was given permission to relocate on three different Indian reservations—the Colville

Reservation northwest of Spokane, the Umatilla Indian Reservation in Oregon, and the Lapwai, Idaho, Reservation. The latter was where Joe's father was born. Young Chief Joseph and his band never saw their beloved valley again.

Joe took the Homeland Project to heart and was soon deeply involved. His gentle and amicable nature and quietly assured manner became recognized as leadership qualities. He was made vice president of the group, later becoming its president. In a year's time, in 1996, the organization created a nonprofit called The Wallowa Band Nez Perce Trail Interpretive Center, Inc.

The organization set three goals to carry out its mission. The first was to find and purchase property, so the powwow didn't have to be held in a gymnasium or on borrowed land. The second was to create an arbor for dancing and drumming and to develop an interpretive center to educate all peoples about Nez Perce culture. The third goal, one that would not happen for nearly two decades, was to design and build a longhouse that could host the traditional Friendship Feast.

The most pressing need at first, as Joe says, was the hunt for land and funding to pay for it. "We had a couple of false starts on properties. Unfortunately, once people found out we were a looking for lands for Indians, willing sellers all of a sudden weren't so willing. Finally we found a person who said he would sell us 425 acres just outside of the town of Wallowa."

It is a beautiful spot, with Wallowa River frontage, golden rim rocks, and views of the majestic Wallowa Mountains. It was also the place of traditional Nez Perce summer camps and grazing areas.

"Of course we couldn't buy it all at once," Joe continues. "We didn't have the money."

Once they made the decision to try to acquire it, donations began to arrive. Where the bulk of the money would come from, however, was uncertain. Time was running out. Then, fortuitously, something happened to give the project the boost it needed. Remarkably, as Joe details, it came from an eloquent appeal made by a national park ranger, Paul Henderson, during a meeting at the Oregon Trail Interpretive Center in Baker City.

Henderson was particularly committed to the project and was aware that on the agenda was a discussion of funding for park interpretive sites in Oregon. Attendees were evaluating the best ways to use monies that had become available from the sale of license plates for the Oregon Trail Bicentennial and from other willing foundations. During the conversation, Henderson raised his hand.

"I'm here representing the Wallowa Coalition of the Nez Perce Trail Interpretive Center," he spoke. "In all your talk tonight, you are forgetting about a trail. There was an Oregon Trail going into Oregon, sure. But there was also another one. There was a trail that left Oregon. The Nez Perce Trail.'"

Silence enveloped the room. The conversation then took a turn. People started to dialogue, to talk about making something work. Smiling, Joe relates the conclusion of that meeting.

"The Oregon Community Foundation awarded us $250,000. Then more donations started coming in from folks throughout the county. From those we were able to afford the first 160 acres of our property."

Later, the Homeland Project acquired more acreage, doubling the site to 320 acres. Today, the arbor has been built, where people drum and dance at the annual festival. Trails wind across the rim rock hilltop, constructed with the assistance of the Nez Perce Trail Foundation, National Park Service, and US Forest Service. The band has an interpretive center, currently located in downtown Wallowa, and informative signs and plaques are placed along the scenic road that leads from the entrance of the property to the Wallowa River.

Raising money to build the longhouse was the most time-consuming project. Completed in 2016, it is central to the Nez Perce Wallowa Homeland Project, and where the Friendship Feast is held. Joe, in his signature quiet tone, shares what it means to him.

"After the Nez Perce war, our people were displaced onto three reservations. There was not any one thing that brought them together. There were no celebrations or powwows on any of the other reservations. I felt strongly that this place could be that—where all of our Nez Perce family could come together and celebrate once again, where after a long absence the drumbeat and the dancers and the songs could be heard from the peaks of the mountains to the deepest canyons of our homeland."

The Tamkaliks celebration, as it is called, happens every July and runs for three days. Traditional dances and drumming are held in the arbor. On the third day, the Friendship Feast occurs, and up to eight hundred people attend. Joe brings and cooks seventeen salmon, while others provide an elk and a buffalo from a local rancher, and the rest of the community contribute a big potluck.

"What makes it special is that everything is open to everybody," says Joe. "People come from all around, and many camp. Everyone is welcome to participate in the services. One happy thing for me is to see that some of those people come from families who were old Indian fighters."

Joe takes great pleasure in the success of the Nez Perce Wallowa Homeland Project, having remained its president for the last twenty years. He is deeply gratified about the results of his work as fisheries manager for the Nez Perce Tribe on rivers in Wallowa County, as well. The joint efforts of tribal biologists and the Oregon Department of Fish and Wildlife have proved exceptionally rewarding. "We work well together and are good friends. The tribe concentrates primarily on helping to restore migratory fish, the salmon and steelhead. The resident fish are managed by ODFW. Together we work on kokanee salmon restoration."

Numbers of native salmon have made great strides. At a weir on the Lostine River, on land managed by the tribe, over two thousand chinook salmon have returned to the river at a location where in 1999 only thirteen were observed. In fall 2018, coho salmon returned to the Lostine River for the first time in nearly forty years. In terms of meaningful outreach, Joe believes this is a first time in America where a tribe has come together with a county to work on the Endangered Species Act.

Most recently, Joe has been appointed to the board of the Grand Ronde Model Watershed, which reviews projects to receive Bonneville Power Administration funds assisting ranchers and farmers to restore rivers and streams. The position gives him one more way to advocate for fish.

"A spiritual elder once said, 'When we have healthy fish, then we will have healthy deer and elk, healthy birds and plants, and a healthy climate for ranching, farming, and logging," says Joe thoughtfully. "With our global need for water, conserving our watersheds is crucial. The water, I believe, should go to the fish."

In addition to his fisheries work, Joe continues to be engaged with local conservation groups. His induction to land-use planning came in 2004, when he learned that sacred Nez Perce ancestral land on the shores of Wallowa Lake, next to the burial site of Old Chief Joseph, was slated to be developed into condominiums and home sites. Deeply concerned, Joe got involved. He heard about the Wallowa Land Trust, a local nonprofit organization just being formed. It was composed of Wallowa County citizens who sought to protect the rural nature of Wallowa County. Joe met with them and was asked to become a founding member. Through the work of the Wallowa Land Trust, what is now the celebrated Iwetemlaykin State Heritage Site was preserved.

"In this instance, we were brought together in our attempt to get federal recognition for the property. We worked hard, and were successful," says Joe. Resulting from donations from the Oregon State Parks Trust, the Nez Perce Tribe, the Confederated Tribes of the Umatilla Indian Reservation, the Confederated Tribes of the Colville Reservation, Oregon Parks and Recreation Department and other grants, $4.1 million dollars was raised to purchase sixty-two acres of the scenic land. In 2009, the property was dedicated. Iwetemlaykin is a Nez Perce name that translates to "at the edge of the lake." The spiritual ground and scenic parkland on the shores of Wallowa Lake and the foot of the towering Wallowa Mountains is now visited by a quarter of a million people each year.

Joe sometimes quotes Young Chief Joseph, who said before his death in 1904, "I love that land more than all the rest of the world." That is the same feeling that Joe has for his beloved Wallowa Valley, and the reason he continues to work to defend it and to protect the spirit of Tamkaliks. He confides that a spiritual elder once told him that he was thankful to the people who occupy the land here now for what they are doing to keep it a beautiful place. In his quiet, humble, and steadfast way, Joe McCormack continues to play a large part in helping it remain that way.

Today there are three trails integral to Wallowa County. One that had been used by early settlers coming to Oregon across the plains in their covered wagons, seeking a better place. One where a first nation was forced to leave its homeland.

Now there is a third trail.

For the Nez Perce, it is a trail leading back home.

Joe McCormack is one who has returned home, to the land of his ancestors. He brings with him friendship, understanding, and forgiveness. He is proud to say that, at the Nez Perce Wallowa Homeland Center, you will never find a no trespassing sign. All are welcome.

Of course it would be that way. For like his father before him, in Joe's home there is always room for everyone's feet under his table.

We are different from other places. Oregonians have a sense of wanting to protect what we have, because there is so much beauty. We have a history of defending our resources. We have a tradition of citizen involvement. That, to me, is part of Oregon.

———————

Jeanne Roy

Perseverance is everything. If you look at people who really make a difference, you see they go for it. In a time where there is so much bad news, this takes on even greater significance. The important thing? Act as if you will make a difference.

———————

Dick Roy

10

A Reverence for the Earth
DICK AND JEANNE ROY

When a husband and wife team from Portland created the Northwest Earth Institute, their mission was to motivate people to take responsibility for the earth and to attempt to live toward a sustainable future. In twenty-five years, their focus has not changed, but the impact of their classes has grown exponentially. Today, over 165,000 engaged citizens from all fifty states and British Columbia have participated in their creatively designed courses, producing positive changes in Oregon and communities throughout the nation.

∽

Two major turning points in Dick and Jeanne Roys' lives both happened on retreats. Each would result in a future neither one could have predicted.

The first occurred at an Oregon State University Student Government Leadership gathering when Dick, senior class president, met Jeanne, humanities senator. That encounter sparked the beginning of what would become a lifelong, devoted marriage and partnership. The second happened years later, when Dick and Jeanne were on sabbatical at the Oregon Coast, examining their priorities and reflecting on their future. What they concluded would, over a quarter of a century, touch countless Oregonians in a very personal way, and from there, reach out across the nation.

Natives of Portland, Dick and Jeanne each grew up with rich exposure to the natural world, which they both instantly gravitated to. With their families, they hiked and camped throughout all the corners of Oregon. As a child, Jeanne developed a special affection for the Columbia River Gorge, the Cascade Mountains, and the deep hemlock forests of Oregon's Coast

Range. Dick's attachment to wild places began in eastern Oregon, while living in Baker City from age twelve to eighteen.

It was there, while on a camping trip with the Boy Scouts in the Wallowa Mountains, that he discovered what would become a lifelong reverence for the earth. That night, when thirteen years old, Dick remembers witnessing Eagle Cap Mountain bathed in moonlight. The majestic 9,572-foot peak, one of the highest in the Eagle Cap Wilderness, was illuminated by a full moon. The shimmering light outlined its broad, bare granitic summit and dark fringe of forest below, where the mountain cradled numerous high alpine lakes. The vision of such beauty and grandeur stirred in him a feeling of awe and gratitude stronger than he had ever known. It was also the beginnings within him of a sense of purpose.

At college, Dick studied civil engineering. He also met Jeanne. The two quickly discovered they had much in common. A week after Jeanne's graduation, they married. After serving time in the military, Dick continued his studies, achieving a master's in structural engineering from Stanford University. He worked for a short time in an engineering firm before he decided that engineering wasn't his true forte.

With Jeanne's support, he changed course and returned to school. He earned his law degree from Harvard Law School. The couple then returned to Oregon, purchased a house, and started a family. Dick went to work at a fine law firm in Portland. By the time they were in their mid-thirties, the Roys had achieved, by most considerations, the American Dream. Yet something, for both of them, nagged.

From the time they met, Dick and Jeanne each held a strong sense of service with a deep desire to live with meaning. Shortly before moving back to Portland from Boston in 1970, Jeanne's internal search for a purpose to her life suddenly became clarified from a remarkably simple event: viewing a photograph of the earth—from space.

"That image changed everything. Humans had just traveled to the moon. We could, for the first time, actually visualize the earth itself, as an entity," says Jeanne. "Seeing the earth from a distance widened and changed my perspective on things. It caused me to realize what a thin layer of atmosphere, water, and soil really sustains all of life. It was the first time it really hit me: humans actually could destroy that which sustains us."

From that moment forward, Jeanne began contemplating the impact that people had on the land. Americans cut down forests, created dams, and paved the landscape—all to produce what they considered was necessary for their lifestyles. "The more I thought about it, the more it astounded me that we could be the cause of so much destruction," she says.

In Portland, Jeanne's concerns only grew. Finally, she knew she must take some sort of action on her principles. She talked with Dick. "I told him I wished to change our lifestyles so as not ruin the earth, but instead, work to protect it." Dick understood; moreover he held similar feelings about the urgent need to try to safeguard the planet.

Busy with three children and a household, Jeanne began to creatively carve out time to work to improve the air quality of Oregon. The Clean Air Act had been enacted in 1963, with a major amendment passed in 1970. Oregon was at the point where it needed to develop a plan to show compliance. Jeanne volunteered to be on the Citizen Advisory Committee for the Department of Environmental Quality.

What quickly caught Jeanne's attention was a debate being waged whether to prohibit backyard burning in the city of Portland. She felt that it should be banned, as it significantly contributed to urban pollution. Jeanne became actively involved, and after six years of intense efforts, backyard burning was prohibited.

Jeanne's next goal was to bring curbside recycling to Portland. At the time, California was a national leader in salvaging programs. Taking cues from that model, Jeanne started an organization called Recycling Advocates.

"I thought, 'Why can't we do this here?'" says Jeanne. "So, we advocated. We took pictures of curbside recycling to show at City Hall. We gave them photos of the bins, of how the materials were collected. I created the Portland Master Recycler Program, which later was transferred to the City of Portland and has been ongoing for the past twenty-five years, and in 1987, Dick and I cofounded Recycling Advocates—a citizens' organization to promote waste reduction, recycling, and composting. We were successful! Portland got its curbside recycling."

Dick became involved in environmental efforts in other ways. He served on the boards of the Oregon Environmental Council, 1000 Friends of Oregon, the State Water Policy Review Board, and the Oregon Parks Foundation. He was appointed to the Oregon State Board of Forestry by Governor Neil Goldschmidt in 1987 and reappointed in 1991 by Governor Barbara Roberts.

For all Dick and Jeanne's successful attempts at advocacy, however, they continued to feel—on a deeper level—it was not enough. What bothered them was something they both observed. They discerned that people, while going through the motions in their everyday lives, seemed to lack a sense of the preciousness of the earth. There was little feeling of reverence or awe for the gift of the biosphere. This problem of perception was of compelling importance to Dick and Jeanne,.

"In the global economy, every place is viewed as an economic opportunity," says Dick. "Many businesses aren't local; they come and go. Everything is objectified. The sense of place is greatly diminished. Too often, people look at the earth as merely a commodity, and this worries me. I think that as long as you view your place only as an economic opportunity, you will probably destroy it."

The deepening apprehension they both felt propelled them to make a profound decision that would impact both of their lives. In 1991, Dick and Jeanne took a four-month sabbatical on the Oregon Coast. Their intent, which they shared only with each other, was to contemplate their life's purpose for the remaining time they had together.

On examination, the facts were fairly straightforward. They did not have a lifestyle that required much money. They didn't travel. Neither needed a lot of 'stuff.' By saving, they had enough to get by. By living frugally and sustainably, they could make it last. They also concluded that their highest vision was to make the best possible contributions to their community. At present, they were conscientiously giving money to causes they believed in.

But there was still a problem. Dick and Jeanne both felt they wanted to do more. They recognized that they had been born at a fortuitous time in history. They had found each other. And the earth they cared for was in danger.

As a couple, they made their choice. Dick would resign from his law firm. They would change their lives. It was a simple but radical alteration.

Instead of giving their *money* away, they would give their *time* away.

When he returned from the sabbatical, to the surprise of his law firm, Dick informed his partners he would be leaving to become a full-time volunteer with Jeanne. He explained that they planned to start a nonprofit to help people see their integral connections to the earth, and to consider how one's actions, however small, affect the natural world and other people. They hoped to encourage the practice of "engaged simplicity"—living fully in place and taking responsibility for that place.

In Jeanne's words, "We are part of the web of life. We are not separate. Everything we do has consequences."

It was how they would try to accomplish that goal—working toward a sustainable culture reflecting respect for the earth and native species—that was revolutionary. The Roys had thought that out, too. They would develop discussion courses to increase people's awareness of vital environmental issues and to motivate groups of citizens to become true leaders for a sustainable future.

In 1993, they founded the Northwest Earth Institute. By some accounts, it was a precarious venture, but Dick and Jeanne carefully considered the process. The sites for their earth-focused education would be new, innovative, and never before attempted: the workplace. In conference rooms and other spaces, groups would assemble to take a discussion course, designed with readings and well-framed questions, leading to thoughtful conversations. Courses would involve seven or eight sessions, each held for one hour at noon, limited to twelve persons. The groups would include cross sections of people—from clerks to executives—all coming together to talk.

"When they are in that room, everyone is equal," says Dick. "There are no right or wrong answers. It's all about the exploration."

"It's entirely different from an advocacy group that's trying to persuade people to do something or to think a certain way. It is based on discussion," adds Jeanne. "Anybody can come in; it doesn't matter what your political views are; there is no judgement. You read a piece, and then talk about it, exploring your reactions and feelings about it."

The Roys tested their pilot course, Deep Ecology, at Dick's law firm. The course was self-facilitated, which would remain the institute's long-standing template. Initially, Dick led the first session. That was the model, he explained. The remaining sessions the group would orchestrate themselves. The final meeting, Dick returned to ask for feedback. The response to the class was unanimous. The participants loved it.

Dick and Jeanne knew they had struck a chord. The next step was to begin knocking on doors to let other companies know about the discussion courses. Because Deep Ecology was self-actualized after the first session, it allowed Dick and Jeanne the flexibility to rapidly get another one going. In time, employees in 325 separate businesses had organized discussion course groups at their workplaces.

Results continued to be overwhelmingly positive. Those who attended the self-examining discussions instigated by the Roys found them provoking and compelling. Partakers admitted the course helped them appreciate nature around them, get in touch with their own values, and find others in the workplace who shared them.

Before long, employees in Portland began asking Dick if they could take Deep Ecology to other offices in the Northwest. Discussion groups expanded by word of mouth from the workplace into churches and community centers throughout the state and outside of Oregon. Three years after launching the Northwest Earth Institute, the Roys held their first national training. Thirty people attended from thirteen states.

Building upon their success with Deep Ecology, they developed three more courses: Voluntary Simplicity, Discovering a Sense of Place, and Choices for Sustainable Living. Each curriculum focused on a different topic, but always with the strong component of a group setting, our connection with nature, and thoughtful, enriching discussions.

Since its launch in 1993, Northwest Earth Institute discussion courses have achieved astounding success, surprising even Dick and Jeanne. The model continues to move organically across the United States. Twenty-five years later, cumulative enrollment in the classes has exceeded over 165,000 persons, with participants from all fifty states and British Columbia. Through it all, the direction of the classes continues with a precise mission: helping people understand the value of consciously shifting passions from activities that consume resources and energy to those that do not.

In 2006, thirteen years after founding the Northwest Earth Institute, Dick and Jeanne's life took another path. After much deliberation, they decided to turn the organization they had thoughtfully created, grown, and nurtured over to paid leadership to maintain its vision for the future. The institute was financially sound, employed a staff of ten, and had a solid prospectus for successful continuation. The Roys determined it was time to dive deeper into the trenches to inspire direct citizen action to build a sustainable future.

In 2007 Dick and Jeanne established the Center for Earth Leadership. Its goal is to motivate individuals to act as agents of change within their own circles of influence and to model ways of living that do not put undue stress on the natural world. The Roys developed distinct trainings for their new project. Participants who attend agree to select a circle in which they are connected, to develop a plan, and to act as an agent of change—for their neighborhood, for their place of work, for their church.

Dick and Jeanne wanted to reach individuals who had a real desire to do something positive and tangible and make a difference for the earth. They wanted to make people think about their behavior and principles and to impact how they choose to live their lives, with actions that are truly sustainable. In its first ten years, the Center for Earth Leadership has trained over twelve hundred people to act as agents of change within their communities. That number continues to grow, and the ways people are changing their communities are varied and amazing.

One person convinced the City of Lake Oswego to create a Climate Action Plan. Another had solar panels installed on her church. One person initiated a quarterly community recycling event, and another person, after completing the course, started a NE Portland Tool Library. Still another agent of change successfully advocated for Pickathon, a large three-day music event held in Portland, to replace all their disposables with washable bowls, cups, and utensils.

"People are becoming increasingly aware that they can't rely on someone else to do what needs to be done. Public agencies just aren't capable of it. So, we emphasize that we better do something from the ground up," says Dick.

A good place to start? Jeanne is unequivocal on that. *Oregon.*

"I think we are different from other places," Jeanne reflects. "People who are born here, and those who are attracted to come here, love to get out into nature. Oregonians have a sense of wanting to protect what we have, because there is so much beauty. We have a history of defending our resources. We have a tradition of citizen involvement. That, to me, is part of Oregon."

"We feel blessed to be part of that tradition," adds Dick.

"We are the opposite of entitlement," says Jeanne, smiling.

Both she and Dick know that their undertaking often runs counter to present-day popular living. But that does not deter them. Their mission is stronger than that. "What we would like to see is a powerful sense of the earth getting into peoples' consciousness in a highly spiritual way," says Dick, "and with that, a robust sense of place. The kind of place where people love their immediate neighborhoods; feel a sense of a village; and enjoy doing things with other people around them by maximizing their interpersonal relationships, while minimizing the time spent in the electronic world."

Within that message is a call for a renewed attachment to nature. It is something that both he and Jeanne fear we are losing and part of the underlying force that drives them to continue their purpose with steadfastness.

"Perseverance is everything. If you look at people who really make a difference, you see they go for it. In a time where there is so much bad news, this takes on even greater significance," Dick says. "The important thing? Act as if you will make a difference."

In their joint years of sustained advocacy, one thing has made a huge difference, they confess. In all their pursuits, it has helped them to endure, to hope, and believe there can be a better future for the earth. Working side by side.

Dick looks at Jeanne with a mixture of love and admiration. "It has been really wonderful to be able to do it together. We are unique in that we have exactly the same vision and goals. We feel profound gratitude for the natural world and the joy it provides."

Jeanne nods and returns the glance, with a smile for her husband.

"Yes," she says. "We were made for each other."

Why do we have these special places?
It's because we have been caregivers of Oregon.
That is the thing that makes Oregon different.
We have remembered that much of our land
belongs to the public, and we've treated it as a
public asset. We have stood up for them. We have
a Columbia River Gorge Scenic Area and public
beaches that are unique in the entire country
because of what individuals in Oregon have done.

Barbara Roberts

11

A Governor for Seven Generations
BARBARA ROBERTS

For her resolute defense of the Endangered Species Act and protecting the health of Oregon's forests, she was the only Oregon governor to face recall three times. Withstanding all threats, her courage defined the spirit of Oregon and its continued commitment to sustain its natural resources.

∾

The path to becoming the first woman governor of Oregon, and, at the time, one of only ten ever elected in the nation, was not a journey for the faint of heart. Any one of the many roadblocks along the way could have easily quelled a soul's spirit for public service, regardless of how strong that determination may have initially been. But instead of acquiescing or resigning to defeat when the obstacles arose in Barbara Roberts's life, she did quite the opposite.

With courage and a tenacious positive attitude, Roberts faced challenges and overcame them. Every test only seemed to forge a stronger belief in the importance of looking out for others who needed help, and for protecting the natural heritage and beauty of the state she fiercely loves.

"I have roots," says Barbara, unequivocally. "I grew up understanding I came from pioneer stock. Oregon has always been who I am."

Those roots, a stabilizing force in her life, run deep. "My great-great grandparents came on the Oregon Trail in 1853 with three children. My great-great grandfather Boggs had land grant acreage and purchased land near Roseburg. Boggs Orchards is still there. When I was born, we were living in Corvallis, but we moved and I grew up in Sheridan. It is Yamhill County I consider as my hometown. I feel a true connection with that town and its beautiful valley. It was there I went to grade school and high school. I was married in that county, six months before I graduated from high school in 1955. I was eighteen."

So began the first challenge.

The decision to marry so swiftly and so young came from the understanding that her husband, who was in the military, was going to be sent to Europe for several years. The only way Barbara could accompany him was if she were his spouse. It didn't happen that way, however. Rather, his first assignment was in Texas. After they married, Barbara spent three and a half years in Texas, coming home to Oregon when his military service was up, bringing with them one child and another on the way.

Just after Barbara turned twenty-one, her second child was born. While her husband attended college, she stayed home to look after the children. Thirsty to learn, she read all his college books and helped him write papers. Eventually, she started taking one class a term at Portland State University. Shortly before her husband's graduation, Barbara's life took a dramatic turn. After sixteen years of marriage, her husband announced he was leaving her for "someone new."

They divorced. Barbara was left with two children, one autistic, and no child support. "This is not the path to the governorship most would recommend," says Barbara, sagely.

What happened next in Barbara's life would forge a permanent mark on the future of Oregon.

"When my autistic son started first grade in the Portland School District, he quickly got sent home because he didn't fit into the classroom. At the time, in the 1960s, there were no laws—neither state nor federal laws, no protections of any kind—so these kids had no way to go to school. I was working as a bookkeeper at a little construction company in Sellwood to make ends meet, but made time to travel to Salem one day a week to talk to my state legislator, Frank Roberts. He mentored me, helping me understand what I needed to do. We organized a group of parents who wanted to get a law passed that said our kids could go to school. Frank Roberts then got a bill drafted in 1970."

The bill, SB 699, would require the state of Oregon to provide funding and support for children with emotional disabilities to attend public school. The state legislator had explained to Barbara just what it would take to get it passed.

"'Somebody is going to have to lobby this,' Frank told me. 'You can't just put it in the legislature and expect it to pass.'"

That somebody, of course, was Barbara. She had no political experience, but the other parents decided that she was the articulate one and should go to Salem. Barbara worked out a deal with her boss at the construction company: every Friday for six months she would travel to Salem and lobby the bill through the legislature, testify before committees, and meet with legislators.

"At the end of the legislative session in 1971 we were successful in passing the bill!" relates Barbara, triumphantly. "It was the first law of its kind in the United States. Oregon now had a law requiring special education for 'emotionally handicapped' children. Nearly six years later, the federal government passed the act that required education for children with disabilities. Oregon was five years ahead of the curve."

From that experience, Barbara realized she had just done something extraordinary: helped to enact change that meant her son now had the right to an education. Further, she had loved everything about the legislative process. It gave her a new way of thinking about herself.

Through the process, she and Frank, both single, had become close friends. The friendship developed into a romance. Then it turned into a marriage. Barbara laughs.

"So out of the deal of Senate Bill 699, my son Mike got an education and I got a husband!"

It also provided Barbara with a helpmate who wanted her to become all that she could be. Frank believed in her and wanted her to go to school. "He told me, 'You are smart, you really should do this.' And it was something I truly wanted."

Once again, the route was complicated. Frank's role in the legislature, first as state representative and then state senator, required unrelenting trips between Salem and Portland. That didn't stop Barbara though, and while fitting in classes at Portland State University, she also continued to be involved with government. She ran for and served on the local Parkrose school board for ten years. She sat on the board of Mount Hood Community College for four years, and then, in 1978, was appointed to the Multnomah County Board of Commissioners, where she became increasingly involved with land-use issues—approving the first light-rail project in Portland between Portland and Gresham and helping design the Multnomah County urban growth boundary.

The urban growth boundary (UGB) policy had recently been implemented throughout Oregon as a result of Governor Tom McCall's signing of the visionary land-use bill, Senate Bill 100. The landmark legislation, which became law in 1973, had been composed by a remarkable coalition—farmers working with environmentalists and supported by far-seeing politicians. Senate Bill 100 requires that every city and county in Oregon set long-range planning goals to address future growth, attempt to control urban sprawl, prevent loss of prime forest and agricultural lands, conserve natural resources, and assure that the state's natural beauty be protected.

Oregon's UGB policy, as part of the larger set of statewide land use long-range planning goals, was another first for Oregon. Moreover, it was the first statewide urban growth boundary policy to be enacted in the nation. The urban growth boundary was a regional delineation. It mandated areas inside and outside of the boundary. Land inside could be used for urban development. Areas outside the boundary were to be preserved in their natural state or used for agriculture and forestry purposes.

Used by local governments as a guide to planning and zoning, the UGB circumscribes an entire urbanized area. It was also something that Barbara strongly believed in. Working in her position as a county commissioner, its implementation was fully concordant with her own values and feelings for Oregon, that the state's natural beauty, native resources, and agricultural bounty needed tools for their protection.

Barbara's interest in politics and passion to serve Oregon continued to grow. In 1980, she ran for the state legislature and won. In 1982 she won again. The following year she became the Majority Leader of the House—the first woman in Oregon to ever hold that position.

In these roles, Barbara saw she could make a difference. In 1984, she entered her first statewide race—for Secretary of State. It was a consequential position and a very hard race that required endless campaigning. Frank was supportive; her children were grown. She toured every "nook and cranny" of Oregon. Again, Barbara won. She was the second woman ever elected as Oregon's Secretary of State and the first Democrat to hold that office in 110 years.

"I took a path in my personal life that should never have led me toward the governorship, ever," she relates of her journey. "Consider the facts: my early marriage, being young mother, having a child with a handicap, and divorced. When I ran for the Parkrose School Board the first time, I had people tell me they wouldn't vote for me because I was divorced and that was a bad example for the children!"

But success continued to follow her, regardless. In 1988, Barbara was reelected Secretary of State. In 1989, she won a fellowship at Harvard to go to the Kennedy School of Government for three weeks to study leadership, an experience that transformed her thinking.

"My brain was full of new and exciting ideas with theories and tools for leadership. I also began to understand that this class of leaders around me had seen me as a leader. When I came back to Oregon, to the Secretary of State's office—the second highest elected position in state government—I understood, for the first time in my life, I *was* a leader."

In 1990, Barbara felt ready to take on her greatest challenge yet—to run for governor of Oregon. She knew state government, knew state people, and she was finishing up all her academic credentials. She acknowledged, though, it was a contest she wasn't supposed to win.

"When I started the race, the first polling showed I was 28 percentage points behind my opponent, Dave Frohnmayer. Dave had already well over $1 million in his campaign account. I had $12,000. Plus, Oregon had never had a woman governor."

As if that weren't enough to test Barbara's resolve, in the midst of the campaign, the spotted owl was listed as a federal endangered species. The law called for protecting "critical habitat"—natural landscape surrounding breeding and feeding sites of endangered species. That meant that some timber lands would become off limits for logging. The listing resulted in unprecedented agitation in the forest industry. Barbara, true to her convictions, supported the Endangered Species Act, and did so publicly. Her opponent took the opposite position, promising that he would fight the Endangered Species Act all the way to Washington, DC, and would work tirelessly to "delist" the spotted owl. The campaign quickly became focused on that single issue—the spotted owl and what the Endangered Species Act would mean to the future of the forest industry.

Barbara sighs. "I was the bad guy during the campaign. Because I lived in Portland, people didn't understand that I'd been born and raised in a timber community. They saw me as 'Portland, Woman, Liberal, Environmentalist.' That was a lot of strikes!" she admits, frankly. "But I also knew we must learn how to manage our timber differently, to use it differently. And I was cognizant of just how the timber industry got where it was. I understood exactly what they'd done over the years."

On the surface the calamity was the spotted owl, but the underlying issue was how we were managing our forests. "The problem was over-harvesting for over one hundred years with minimal replanting for most of that period of time, as well as shipping logs overseas to be processed and not processing them in our own mills. We made every mistake we could to make the industry not perpetual. We were harvesting old-growth timber and making toilet paper out of it! It was all about short-term gain."

While she was loved by the environmental community, Barbara had the support of only one timber company in all of Oregon, based in Eugene. Never in Oregon history had there been a governor elected who was not supported by the timber communities. She continued, however, visiting towns throughout the state and expressing why she held her views. At each meeting, she spoke directly about the science behind her reasoning.

"We have been dealt a tough hand. Now we have to figure out how we're going to meet the letter of the law and get back in the woods, and out of the courtroom and back in the forests of Oregon. We must learn how to manage our timber differently, to use it differently. We will not 'throw in the cards' but must play the hand we're dealt."

While people in the affected communities did not like hearing what Barbara was saying, many came to understand the real issues facing them. They could see in her a genuine empathy for the hard times facing them. They grew to respect her courage, and many came to see she was not condoning that that timber was not to be used; rather, Oregon must begin using its resource more wisely than it had done in the last hundred years.

In November 1990, the state voted. Barbara Roberts accomplished something no one had done before. She became the first woman governor of Oregon.

Objectivity, framed in science, would be the hallmark of Barbara's environmental stand for Oregon. Her endeavors were overwhelmingly successful. Part of the reason was that, unlike many politicians, she was never afraid to lose; winning at all costs was never her goal. Oregonians could see that. Her goal was winning for Oregon, and saving for future generations a place she deeply loved.

Barbara reflects more about her passion for Oregon. "Clearly my parent's love for Oregon was a foundation. I have always felt a real historical connection because my family came on the Oregon Trail. My parents took me fly-fishing and camping in Oregon. I didn't really want to fish, but I could sit by a stream and watch water going over rocks for hours. I've never lost that love of water. When I was campaigning, I would make my staff stop when we went on the Santiam Pass so I could go sit for fifteen minutes and just watch the river."

As governor, Barbara's resolve to protect what she valued about Oregon would be tested many times. Early in her tenure, several significant land-use issues faced the state. Many were contentious. Some could have changed the shape of Oregon.

"Not on my watch," intones Barbara, resolutely.

Preventing a hydroelectric facility that would have destroyed the last free-flowing part of the Klamath River in southern Oregon is one achievement in which she takes pride. The Salt Caves Hydro Project, if constructed, would have destroyed a beautiful river, its fisheries, and Native American burial grounds. "In addition, the electricity was going to be sold to California. The project was unnecessary."

With characteristic tenacity, Barbara worked to get a bill to make the Klamath River a state wild and scenic river onto the statewide ballot. She was successful; the measure passed. But having the Klamath become federally designated as such was even harder. Barbara had to lobby and prove to the Secretary of the Interior that it was worthy under the National Wild and Scenic Rivers Act.

It was a long process, but one she would not let go, even in the final month of her term as governor with scores of items demanding her attention. Barbara insisted on continuing pressure until the designation was made. Finally, just before leaving office, the Klamath River was designated as a federal wild and scenic river. The construction threat to the river was put to an end.

Barbara also worked to pass a progressive mining law for Oregon. How she accomplished it, however, was unconventional. "I arranged many different people—those who had mining operations, those who were environmentalists, and all kinds of engineers—all together in the same room in the back of the governor's conference office," she says, with a grin. "We worked for four straight months. We came out with a bill all sides agreed to. We got it passed." Not only did they get it passed, it was, and still is, the strongest mining law in the country.

Another environmental hurdle that concerned Barbara was the construction of the Elk Creek Dam in Jackson County. If built as proposed, biologists warned it could have disastrous consequences for native fish runs. Barbara took the matter to heart. She lobbied fiercely in Washington, DC, for dam removal to protect native fish populations. She was successful. Elk Creek Dam was never completed.

Possibly the most far-reaching environmental threat during Barbara's office as governor occurred during the 1991 and 1993 legislative sessions, when she faced severe pressure to dramatically weaken Oregon's emblem land-use laws. If she did not acquiesce to the demands of those in the legislature, she was duly warned, they vowed to "zero out" the budget for the Land Conservation and Development Commission (LCDC). This was a serious threat. LCDC had been established with the passage of Senate Bill 100 to oversee the mandated land-use planning program and to assure compliance of local governments with the goals of the law.

The pressure was intense. The senate was adamant about weakening the laws and the statutes that protect forests and farmlands. Barbara did not flinch. She knew that if she allowed it to happen, the land-use watchdog would be less effective. Far better, she reasoned, to have a temporary budget shortfall than a wholesale gutting of the laws undergirding the land-use system. Barbara knew Oregon would never be able to put together a coalition as it had in 1969, 1970, and 1971, to get Senate Bill 100 passed.

A month of intense negotiations ensued between LCDC staff, local governments, and legislators. Ultimately, the legislature came up with a bill that allowed for some fine tuning of the forest protections without undercutting the land-use laws. The basic system was saved.

"My veto pen was always ready to kill any anti-land-use legislation that reached my desk," says Barbara. "I have served on the board of 1000 Friends of Oregon. As a county commissioner, as a legislator, and as governor, I have worked to preserve our land-use laws. I am a firm believer that so much of the beauty we have today, and so much of the economic opportunity such as our vineyards are there because we didn't fill them up with condominiums and shopping centers."

Where do Barbara Roberts's convictions originate? They arise, she explains, from an intrinsic understanding of what Oregon truly is.

"Whether you're driving down the Columbia River Gorge, or the Oregon Coast, or through the deserts of eastern Oregon, our natural resources are magnificent. We are all so lucky to live here. I don't need to go to Europe to find beauty or to Asia. I've been to those places. There's enough beauty in this state to last me forever.

"There is not a part of Oregon I don't know, and there is not a part that I don't love. The more I campaigned statewide for my races, the more I fell in love with new places I hadn't seen before. Every time I see us taking care of Oregon—knowing that we have reserved it for the future, that we are making sure that there will be parks and open spaces so that generations from now can walk and picnic there—means we are doing a good job.

"Why do we have these places?" Barbara asks pointedly. "It's because we have been caregivers of Oregon. *That* is the thing that makes Oregon different. We have remembered that much of our land belongs to the public, and we've treated it as a public asset. We have stood up for them. We have a Columbia River Gorge Scenic Area and public beaches that are unique in the entire country because of what individuals in Oregon have done."

Barbara grimaces. "And all this didn't happen by accident. I have the scar tissue to prove it."

Barbara's continued backing of the federal Endangered Species Act and the listing of the northern spotted owl, and her pivotal role in the spotted owl recovery plan, did not go unnoticed by the timber companies. They observed her involvement with the Clinton Forest Plan and took it as a lack of support for the forest industry. She was perceived by them as one who was "locking up the forests," "throwing Oregonians out of work," "willing to raise taxes rather than cut government spending." To many of those in the business, this was unconscionable. It could not go unchallenged.

In March 1992, a petition was circulated across the state to recall Barbara Roberts because of her environmental positions. It was the first time in Oregon history that a recall petition was filed against an acting governor. The recall failed. Challengers were not deterred. Funded by lumber magnates, a second recall was attempted. It too failed. Then, in October 1993, a third effort to recall Barbara Roberts as governor was announced. It also met the same fate.

"These three recall efforts against an Oregon governor were, yes, another first, but one I could have done without," says Barbara.

Her courage, however, in standing up for what she believed was important for Oregon proved a lasting legacy. In time, Barbara received numerous awards for her work. She was honored with the Governors Gold Award for Extraordinary Service to Oregon and two national lifetime achievement awards, one from the Center of Policy Alternatives and the other from Women Executives in State Government. The Oregon House and Senate would name a building in the Capitol Mall the "Barbara Roberts Human Services Building." Tributes from environmental groups, children's groups, minority and women's organizations, the American Civil Liberties Union, and disabilities organizations flooded in.

While Barbara feels honored for the recognition, she puts her dedication to the environment in perspective. "We have been good caregivers, not as good as I wish we'd been and not as good as I hope we are going to be, but the point is, we have been caregivers of Oregon. When my great-great-grandparents came as pioneers on the Oregon Trail, they gave me this state as home. What are my great-great-grandchildren going to say that I gave them? If I am making a choice, to build something or not to build it, to cut it or not cut it, to take care of it, preserve it, it's all based on a philosophy that comes from Native American tribes: 'Make every decision for seven generations.' Seven generations from now, I ask myself, would I feel proud of what I did?"

Undoubtedly, Barbara's decisions will have lasting impact for the good of all Oregonians. "What would I like to see happen in Oregon's future?" she continues, reflecting. "I would like to see people who have lived in Oregon for a long time continue to tell their stories and share their place. These aren't our secrets. They belong to every new person who arrives here, to children and families, no matter where they came from. If they came on the Oregon Trail, terrific. If they came from Mexico, great. If they came in on a plane from the Middle East, I'm just as excited. If we tell the stories well, new people will come here and feel the same respect for Oregon that we do."

Why do these people keep coming? "Because we've taken care of our state. That's the reason why I don't want Oregonians to believe that anyone who steps forward with 'I'll create jobs as a solution' holds the answer to Oregon's future. I don't believe they do! They need to care equally about the environment in Oregon. I want them to believe that our environment is as much of an asset as any new company that we bring into our state."

Barbara is clear: if Oregon is to continue leading the country in caring for its natural resources, its spectacular beauty, and its people, it will require all of our efforts. As much as anyone, she understands that this road can be difficult. Still, it is worth the cost.

"How would I like to be remembered?" she asks. "I would hope that people might look at my work and say, 'She took the strong stand. She wasn't afraid to be unpopular if that was what was necessary.' Each generation has but one chance to be judged by future generations. This is our time."

She smiles. "Never do I cease to be amazed at the beauty of Oregon."

Our relationship to the land matters.
If we lose the health of these landscapes,
then we are not going to succeed as a society.
We need to make sure that these landscapes
are more than protected. They need to be
regenerated. They need to keep producing things,
not only for the ecosystem itself, but for the
quality of our lives.

_____

Jack Southworth

12

A Resilient Land

JACK SOUTHWORTH

The award-winning rancher and facilitator for the High Desert Partnership asks the group to sit in a circle, bringing them together to talk. There is something remarkable about the assemblage, however: its diversity. It consists of ranchers and government resource specialists, landowners and interested citizens, timber industry representatives and conservationists, all coming together with the goal of working together. They will engage in a pioneering collaborative process, developed in Oregon, to find common ground and solutions to some of the most challenging issues surrounding Oregon's public lands.

∾

In 1967, when he was twelve years old, Jack Southworth had one of the best days of his life. Driving his dad's 100-horsepower tractor to the hay meadow, Jack put a cable around the last willow standing in a three-mile stretch of the Silvies River outside of Seneca in southeastern Oregon. Then, with his proud father looking on, Jack maneuvered the big machine to yank the tree out of the ground. The land, thought the fourth-generation rancher, would now have nice clean vistas. It would resemble the pictures he'd pored over in *Successful Farming* and *Farm Journal*, with hay meadows square and regular, allowing the hay to be cut and baled in an efficient manner.

"I had never felt so good about anything I'd ever done in my life," says Jack, remembering. "My father wanted grass to grow right to the edge of the water and nothing else. The trouble was," he adds, thoughtfully, "that's not what the river wanted."

At the time, though, removing trees and excess vegetation was considered best possible management practice by the Soil Conservation Service (SCS), a federal agency created in 1935 and renamed in 1994 the Natural Resources Conservation Service (NRCS). Employing soil scientists and other professionals, the SCS provided technical assistance to farmers, ranchers, and private landowners, with the mission to improve and conserve the nation's soil and water resources on private lands. In the 1960s, the scientific consensus advocated that ranchers and farmers draw up plans to straighten rivers so there would be more land to hay and to cultivate.

As he grew older, however, Jack observed the effect these practices had on the landscape. Without adequate vegetative protection, as a consequence of the demise of all the willows that once flourished along the meandering Silvies, the river's banks began to erode. In a desperate attempt to stem the erosion, Jack's father began depositing old cars in the water along the most severely eroded banks. They did little good. By the time Jack left for college in the mid-1970s, the ranch of his youth, while still productive, had little resemblance to what his great-grandfather encountered when arriving to the high desert country in 1885.

"My great grandfather, William, homesteaded the ranch that year. It was a 160-acre operation," says Jack, explaining the history of the ranch that has been in his family for over 133 years. "He had a small saw mill near Canyon City, Oregon, and needed hay for his horses and oxen, so he took out this homestead and cut hay with a hand-scythe. My grandfather, William's son, along with my great uncle, decided they could turn the homestead into a ranch. In 1948, my parents took it over and they continued adding to the ranch. Through all the years, the ranch management resulted in the ecosystem growing simpler and simpler. Productivity was everything, following the 1950s, '60s, and '70s philosophy. Bigger is better, and more is better yet."

In 1978, Jack graduated from Oregon State University. Accompanied by his new bride, Teresa, whom he'd met at school, they returned to the ranch. Jack's parents retired, and he began to manage it. The place was free of debt and still profitable. Over the next few years, Jack and Teresa made some additional land purchases. About the same time Jack and Teresa started running the ranch, however, the national economy did a backflip. Interest rates went up; cattle prices dived down.

"Suddenly, in the early 1980s, Teresa and I found ourselves $1 million in debt, with 15 percent interest rates. It was touch and go; we nearly lost the ranch."

Unsure what the future might hold, they were open to new ways to manage the ranch and heard about a wildlife ecologist, Allan Savory, who was introducing innovative methods to manage rangelands in dry environments similar to theirs. Originally from South Rhodesia, Savory offered a different kind of grazing system, one that protected arid lands from desertification and environmental degradation. He called it "holistic livestock management."

Savory was offering a week-long course in Albuquerque, New Mexico, and Jack and Teresa, faced with losing their operating loan in 1984, decided to take it. "We got in our Volkswagen camper van and drove to New Mexico and spent $2,000 we really didn't have," says Jack. "What Savory was saying was an anathema to me and to all I'd learned about grazing and grazing systems. But Teresa, who hadn't taken any range courses in college and had nothing to unlearn, soaked it up like a sponge."

Jack's father was not happy with their decision to attend. Jack had given him a book written by Allan Savory and he completely disagreed with his approach. "My father referred to it as 'less-than-savory-grazing' and 'holier-than-thou-management.'" But to Jack, as hard as it might be to admit, aspects of the Savory method were intriguing.

"After we came back, we used some of the ideas about economic analyses and grazing, and it enabled us to stay in business. The main thing offered from holistic management is the creation of a three-part goal."

As Jack explains, the first part is to decide on the quality of life that you want. The second is to define what you want to produce on the land. The third is to seek to create the landscape that will bring about those two things.

The discussion about a new form of ranch management took place in the dining room of their home with Ed, their one employee at the time. Writing down everything on paper was probably the most painful thing he'd ever done, says Jack. It got him talking about things—such as "What do you want in life?"—that he'd never spoken of before. It dealt with activities that didn't have strictly a financial return. Profit was still important, but profit was not the only thing. Moreover, it suggested an entirely different methodology to ranching. In the past, ranching meant attempting to control the land. The new approach, which Jack would fully embrace in the years ahead, was to go with what nature gave you.

Jack used a flip chart to map out their goals. First, Jack and Teresa ascertained that what they wished for as a quality of life was to achieve satisfaction and a feeling of self-worth from living and working on the ranch. They wanted to be able to also have a day off now and again, and to be able to participate in the community beyond the ranch. They desired to think of their ranch as an opportunity to achieve their life goals, and not an end in itself.

Their second goal was to have cattle that were born, bred, and raised on the ranch. They wanted a forest with trees of different sizes and ages that were fire tolerant. They wanted streams that were lined with willows, stocked with beaver, and they wanted to create good habitat for fish. They wanted healthy grass plants.

The final goal was to seek to create a landscape that would allow this to happen. That meant their grazing regime had to change and be controlled to allow a dense stand of perennial grasses and some shrubs. They wanted the precipitation they received to stay on the ranch as long as possible and leave as late season stream flows or plant growth.

What they concluded was, while they may not make as much money, they would have a lot more fun doing it. They also realized that if they didn't plan for these goals, "they will slip away from us almost without our realizing." Jack and Teresa also recognized that the change in management would take time and effort and would mean putting back many things that earlier generations had taken out.

"At the time, our cows, grazing right up to the water's edge, were really damaging the creeks. We knew that, to be good managers, we needed to fence off sections of the Silvies River to keep them away from those areas," Jack says.

One day soon after Jack and Ed began the fencing project, Jack's father saw Ed driving out toward the meadow with fence posts in the back of the truck. He asked Jack what he was doing.

"I told my dad, 'I'm not going to tell you. You won't like it,'" says Jack. "My dad replied, 'You've got to tell me.' So I told him. 'We are going to fence off a quarter mile of the Silvies River to keep the cattle away from the creek. We want to see if we can reestablish willows and have a healthier stream bank.'" Jack shakes his head, remembering what his father said next.

"'I spent my whole life taking out the homesteader fences and willows and you're just building them right back up!'"

Jack acknowledged it was true. He knew the idea seemed insane to his father. And he understood where his father was coming from. "Back then, when my father was working, people weren't even aware of what a riparian area was," Jack explains in a gentle tone. "So, don't judge too harshly."

Before long, Jack and Teresa began seeing small successes. They hired additional workers. Continuing to pursue their three-part goal, they worked closely with their employees, making collaborative decisions with their input. They used the flip chart every morning to talk about the day ahead.

"I've used the flip chart for the last twenty-eight years," Jack relates, then smiles. "That's the power of a wood stove in Bear Valley in winter. When it's twenty below, we meet at the wood stove and start the flip chart. No one wants to go outside."

To control the movement of their livestock, Jack continued to build new fence, usually constructing a mile each year. Following the Savory method, they stocked their cattle wisely and always kept them moving to proactively manage for healthier, perennial grass plants. And along both sides of the Silvies River, Jack planted willows.

"People ask us about what we are doing on the ranch, about the fencing, all the time we spend planning with our employees, and the things we do that don't seem to directly increase profit. They say, 'Isn't that kind of a gamble?' I tell them yes it is, but I want to gamble on the side of increased participation. I want to gamble on the side of a higher water table. I want to gamble on the side of increased biodiversity. I want to gamble on the side of happier people. And I want to gamble on the side of more resilient ecosystems."

Over time, stream bank erosion was significantly reduced. The willows, always Jack's pride and joy, took off. They now form healthy stands lining both sides of the river, which meanders gracefully through the farm and thankfully, says Jack, was never straightened as had been recommended years before. All the added vegetative growth has created diversity for the birds and insects and increased stability to the environment.

"We used to think about just one thing—staying in business for another year. Forming the three-part goal empowered us to pursue activities that didn't have just a financial motive. Now, we're just as concerned with our landscape and our quality of life. I can look at our landscape and have a feeling for the health of the grasses, the creeks, and see whether we're going toward or away from our goal."

Jack's cattle are also thriving. His ranch was one of the original fourteen Country Natural Beef members. The farsighted Country Natural Beef Cooperative focuses on raising beef in environmentally sustainable ways with ecological stewardship guidelines. Jack has remained closely involved with the program for the past thirty years, also acting as its chairman.

The Southworth Ranch continues to receive exemplary scores for its operation. The Food Alliance cites that "no herbicide control is used; no weed problems are observed; riparian and upland vegetation areas are improved; willows have been planted along streams benefiting both fish and wildlife, great care is taken to keep stress low for animals; stream sections are fenced, providing complete protection for the riparian vegetation inside; dense perennial vegetation is in adjacent fields; and with good litter cover, rapid water infiltration occurs, minimizing potential erosion."

These ecologically based, high standards for range management benefit not only the cows, the landscape, the ranch itself, but also the public land where Jack grazes his cattle in summer, so hay can be grown on the farm for winter feed. Public land, whether overseen by the US Forest Service, the US Fish and Wildlife Service, or the Bureau of Land Management, is a critical component for many ranchers, who pay to use it for summer grazing. Done properly, grazing can be a tremendous tool for reducing fire danger and helping perennial grasses maintain their vitality.

Jack's summer grazing allotments, primarily heavily timbered uplands and mountain meadows, total over twenty-five thousand acres in the Malheur National Forest. He is a careful steward of public land and manages it closely. He makes sure his cattle are off the grassland while it is still growing so the perennial plants can rest and restore themselves. If a rancher is not doing right by water, salmon, and the environment, then that rancher should not be on the land, he believes.

For his outstanding stewardship, Jack has received multiple awards, including the 2016 Range Rancher of the Year in the United States, the prestigious National Range Management Award from the US Forest Service, the 2009 Conservation Farm Award, and the 2010 Grant County Stock Grower of the Year. He has also been inducted into the Oregon State University Agricultural Hall of Fame.

He is clear in his feelings about all public lands, and what he hopes for them. "I'd like to see our public lands well managed in that they are productive for the community, provide jobs for the community, but are also places that are great for recreating. I would hope to be allies with the agencies who oversee them. We need to manage this resource well, and to create fishing, hunting, camping, solitude, and also to provide some cattle grazing and timber harvest. Let's find a way and use our minds and planning and collaborative skills to have forests and public lands that work for all of us."

Jack has sought to make that happen. He was an early board member of the nonprofit High Desert Partnership, which was founded in 2005 to serve Harney County and surrounding southeastern Oregon. (At 10,228 square miles in size, Harney County is larger in area than six US states; it's the largest county in Oregon and one of the largest in the United States. It is also sparsely populated, with a population of only about 7,700.) The organization was created to bringing diverse interests together—land management agencies, conservation organizations, the timber industry, landowners, ranchers and interested citizens—and to support partners engaged in collaboration to solve challenging issues in their communities. The underlying belief is that if the process is attended to, the outcome will be for the holistic good of all. Consensus can result in successful initiatives to restore and bring about healthy forests and wetlands, to mitigate wildfire, to create positive experiences for county youth, and to promote thriving and resilient communities within the rural area.

Since 2008, Jack has acted as a facilitator for one of four initiatives developed by the High Desert Partnership—the Harney County Restoration Collaborative. The innovative collaborative, convened by the High Desert Partnership with the help of Harney County judge Steve Grasty and the

Nature Conservancy, supports and implements restoration work in Malheur National Forest with the goal of creating fire tolerant and ecologically diverse forest ecosystems. As facilitator, Jack has helped diverse interests find positive consensus on many issues—economic, ecological, and social—that directly affect more than 250,000 acres of forestlands. As a measure of his success, none of these large projects have ever experienced litigation.

For many Harney County community members, the High Desert Partnership is regarded as an alternative to the extreme view that publicly owned lands be taken away from federal government control. In 2016, agitators made worldwide headlines when they forcibly occupied the Malheur National Wildlife Refuge in Burns, Oregon. The occupation, a protest against federally owned lands, lasted six weeks. Leader Ammon Bundy of Idaho proclaimed his group was taking a stand against "government tyranny."

The revolt was intended to sow unrest and create discord and polarization. Inciters hoped for broad community approval. To the extremists' surprise, their tactics failed to win the town's sympathy and overall support. Why? As one rancher, quoted in *High Country News,* eloquently stated, "The High Desert Partnership's methods were what inoculated us from the Bundy disease."

Behind that inoculation was ten years spent building trust, communication, and cooperation between community members, property owners, government agencies, ranchers, farmers, and environmental and special interest groups, who sat down together to find common ground.

Jack remains dedicated to the collaborative process. He views it as the best possible way to find solutions to difficult problems, especially concerning our public lands. People often ask him, "Why do you collaborate?" and "Why do you sit in a circle?" Jack's smile is generous. "I tell them, 'Well, I think there are three options. We can litigate, legislate, or we can collaborate. I don't have money to hire lawyers. I don't have the influence to do my bidding. So that leaves collaboration.'"

His tone becomes more serious. "It's kind of like what Winston Churchill said about democracy. 'Democracy is the worst form of government, unless you compare it to all the others.' That's the way it is with collaboration. It's slow, it's tedious, but there is something about going around in a circle and giving everyone a chance to speak and then asking someone at the end of that circle to strive for a consensus, for something we can all live with. I find that very empowering.

"I believe we can learn how to speak with respect, but also we can learn how to listen. That's the thing about the circle. If you know you are going to speak next and don't have to raise your hand or talk the loudest, you might actually hear what I'm saying. That's the power of sitting in a circle and listening and collaborating."

Whether acting as a paid facilitator or considering the choices he has made on his own ranch, Jack has observed that collaboration and holistic goals produce something critically important: *resilience*. It is something we need more of everywhere, Jack firmly believes, especially during these polarized times.

"Today, we are sharply divided in this country, and exhibit a low threshold. We do not have the emotional wealth to say, 'I can stand back.' There is too much exchanging of blows and not enough turning the other cheek. When we use emotional resilience, we can turn the other cheek, and come back with a more civil response."

Jack recognizes another source of resilience that we need to develop too. It harks back to what Jack learned from pulling out the last willow from the creek when he was a young boy. *Ecological resilience.*

"That willow was probably glad to be put out of its misery," Jack relates. "But it was not a resilient ecosystem. It could not take a heavy flood. It could not provide shade for fish in the summertime. It could not provide willows for beavers to fill a dam. That is what we have been trying to bring back. What we are doing now on our ranch is creating ecological resilience. We have healthier plants that are more resilient, better to handle the cold or drought or heat. We cover the soil with plants so have less bare soil and less evaporation for the little bit of moisture we receive in Eastern Oregon."

Looking out over the beautiful high desert ranch he shares with Teresa, as the wind moves the green summer hay in waves and a soaring marsh hawk scans the ground, Jack's profile is one of satisfaction and contentment. He has faithfully managed this land for over forty years, and it has responded in kind. The willows are full of life. Beavers have returned. Wildlife now abounds throughout the ranch, with habitat for elk and deer, antelope, colorful songbirds, and careening raptors. There are fish in the river, and migratory birds reside for a while in their travels in spring.

"Our relationship to the land matters," Jack reflects. "As individuals, families, and communities, we need healthy landscapes in order to continue as a society. If we lose the health of these landscapes, then we are not going to succeed as a society. We need to make sure that these landscapes are more than protected. Protection may not be enough. They need to be regenerated. They need to keep producing things, not only for the ecosystem itself, but for the quality of our lives."

Jack glances warmly at Teresa and raises the question of why, at their age, with no children, they don't sell the place. It is obviously something the two have discussed. They could live very comfortably, wouldn't have to worry about budgets and feeding cattle when it's thirty below zero. Why should they continue to ranch?

"Because it is so exciting," says Jack, unwaveringly. "It's just so exciting to create a budget that enables you to carry on, and at the same time to create healthier grasslands, create healthier forests, create employees who enjoy working here that have kids going to our school. All that results in a quality of life. It's such a fascinating puzzle to figure out; what could you possibly do that would compete with that? Yes, I could retire and go to Palm Springs, but after about the second week of golfing, what would I do? Trying to make this Oregon ranch work is just the most fascinating thing I can think of in the world."

When I consider Oregon—the largeness
of this state, the beauty of its varied landscapes—
I believe it influences people's character.
People are more open here, more welcoming,
more willing to share. When you live in
an environment like Oregon, it creates
a generous nature.

Zari Santner

13

From the Caspian Sea to City Nature
ZARI SANTNER

She could not do the things she had hoped in her native Iran. Now, in her adopted country, there was still time to make a difference. That longing would inspire her to transform a long-neglected landscape and remnant cinder cone into one of Portland's largest and most treasured parks, and to create a first-of-its-kind program: City Nature.

∾

In Persian, Shahsavar means "Mounted King." To Zari Santner, raised in Iran under the presence of its encircling mountains, Shahsaver also symbolizes the reason for her passion to preserve natural beauty in a place half a world away from where she was born.

At one time a small town on the shores of the Caspian Sea, Shahsavar provided a rich landscape in which Zari grew up from 1947 to 1962. Bordered by the sea on one side and the Alborz, a high-elevation mountain range, on the other, Shahsavar enjoys life-sustaining rainfall unlike much of arid Iran, two-thirds of which is a dry plateau. Mountains trap moisture rising from the waters, making Shahsavar a verdant land, lush, almost subtropical. To Zari, as a girl, it was paradise—a paradise lost after the revolution and wars with Iraq.

From a population of twenty thousand when Zari grew up, the town boomed after the Iran-Iraq War (1980 to 1988) with refugees and people fleeing other parts of Iran. In less than thirty-five years, by 1992, it had expanded to a city of over two hundred thousand. Its growth occurred in the absence of any zoning or planning. Mountains were denuded of trees for fuel during the war years; afterward developers bought up the land, creating villas sprawling across the foothills. The Shahsavar that Zari knew and loved as a child was gone forever.

"It is very, very sad to me," says Zari. "But it makes me more determined than ever to preserve what we have here in Oregon. Because if you are not vigilant, if you do not do the right thing, changes happen. And by the time you notice it, and by the time you say, 'All this can't happen!' it has already happened. It is changed and you cannot reverse it."

Zari's affinity to landscape inspired her to study horticulture at the University of Tehran and later at the Harvard Graduate School of Design. She deeply responded to the natural feeling of a place, and as an antidote to the destruction of the Shahsavar she had loved, Zari wished to develop places of beauty. Her passion was guided by a determination to try to preserve what is left of nature before it is gone.

Finishing her graduate work in landscape architecture at Harvard in 1974, Zari contemplated where she, her husband, and young family might travel to live. Her husband had spent a summer at Reed College in Portland, and Zari knew of Oregon's environmental ethics, which attracted her. They decided on Oregon. Packing up their belongings, they moved to a place which Zari had never seen.

When she arrived, its landscape instantly reminded her of home.

"Oregon is grander, bigger, but still with the same qualities I'd loved about Shahsavar," Zari says. "My attachment to Oregon was immediate. I deeply missed my home, but at this time in life, Iran was in such an upheaval and, given the fact that I was a woman and with the current patronage system in the county, I knew they were not going to allow me to do what I wanted to do for my country. Oregon became my adopted home, and all those things I had desired to do for my home country, I decided I would now do here."

Zari's heightened perspective of the demise of Shahsavar—the rapid loss of its beautiful rice fields, its orange groves, its once abundant tea plantations—revealed to her a lesson she has kept with her all her life.

"When push comes to shove, things can irreversibly change," Zari says thoughtfully. "The most detrimental thing to preserving and conserving what we have is a sense of complacency."

This understanding motivated her to action. In 1982, she began work for the City of Portland in Parks and Recreation—a time, she says, when she was surrounded with bright, talented, well-educated, and idealistic employees. There was enthusiasm in "setting direction" in the city—something that intrigued and excited her. She became aware of a new visualization being proposed for her region—a forty mile nature loop.

Actually, the greenway nature loop idea wasn't new at all. It had originally been recommended years before by the Olmsted brothers, highly regarded landscape architects, in their report to the Portland Park Board in 1903. For decades after John Charles and Frederic Law Olmsted Jr. had penned their visionary report, however, the 40-Mile Loop had sat unimplemented and forgotten. That changed when local activist Barbara Walker discovered the historic Olmsted document in the 1980s. Walker became its champion and voice, and tirelessly began advocating that the plan was a remarkable one and needed to be resurrected. The City of Portland had grown, of course, but the proposal was still valid and farseeing. Walker and others believed there should be a trail to surround the city to connect its many citizens to parks and the natural environment.

One providential day, Zari was invited by parks employee Marlene Salon to join a bus tour of the imaginary loop following a proposed trail system that Barbara Walker and her 40-Mile Loop advocates had roughly sketched out. The outing followed an abandoned railroad track called the Springwater Line in southeast Portland extending all the way to Boring, Oregon, in Clackamas County. They wished to investigate its potential, if it could be secured, to turn into a trail.

One of the stops on the field trip was at the base of Powell Butte, located in the far eastern part of city. Spying the six-hundred-foot volcanic cinder cone butte rising above them, Zari and Marlene came up with a notion that would alter the future of Portland. Zari's face lights up. "Somebody on the trip said, 'The city owns this property; I believe it's the water bureau,'" Zari remembers.

They were correct. The butte's location and height made it a strategic element in the city's water supply system. A series of underground storage reservoirs held water from Portland's Bull Run Watershed that would be delivered to residents.

The group explored some ad hoc trails that had been developed on the butte's hillsides. "Before I'd gone a few hundred yards, my heart broke," says Zari. All the trails spanning out on the butte were deeply rutted from the careless wheels of motorcycles. Some of the runnels were as wide as Zari's living room, and nearly unwalkable. The group continued climbing. Then Zari saw it. At the summit of the butte, surrounded by scores of blighted tracks, was a stupendous view of three majestic Cascade Range mountains: Mount Hood, Mount Adams, and Mount Saint Helens.

From that moment on, Zari could not stop thinking about the potential this beautiful spot, if reclaimed, could offer.

"The view took my breath away! It was an incredible!" she says.

It had grabbed Marlene's heart too. On the way back, they both were lost in imaginings. "'What if Portland Parks took over the management of the area from the water bureau?' we asked ourselves. 'What if we created a master plan and opened Powell Butte for public use? What if we put a stop to this denigration and focused on the butte's great beauty?'"

Zari knew very well the challenges of that idea. The Portland Water Bureau had tried fencing off the property to secure it from vandals. It had not been successful. Illegal camping, drug use, and abuse from motorcyclists were rampant. Yet Zari was undeterred.

"Why not develop some kind of plan that says Powell Butte could be used in conjunction with their underground water reservoirs?" Zari recollects. She posed her idea to the director of the water bureau.

The director's answer to her request was concise and firm. "No."

Zari did not give up. She knew this place was an incredible asset, one being ruined because of bad and narrowly envisioned policy decisions. Plus, she says, the public didn't know anything about it. With the fate of Shahsavar in the back of her mind, Zari continued adjuring the water bureau to rethink the idea. They could work together, she stressed, the bureaus of parks and water, for mutually beneficial results.

At last, in 1985, two years after her first visit to Powell Butte, the water bureau agreed to give it a try. Their compromise was a triumph to Zari. "It was like, oh my God! Perhaps I can help protect this 640 acres of wonderful land, not only for people's enjoyment, but keeping it from destruction!" says Zari, beaming. That revelation, though, also held something more. She saw the success as a catalyst for the restoration and preservation of other spaces of nature for the citizens of Portland.

"It has always been my hope that by saving these places, it will open people's eyes to their splendor, that there are more places to be saved ... that once this is preserved, they will want to preserve others."

Powell Butte became an official nature park in 1990. In Zari's mind, it laid the foundation for a bold conception she continues to hold: the creation of a Forest Park of the east.

As Zari explains, on the west side of the Willamette River lies the fifty-two-hundred-acre urban wilderness park and nature preserve, Forest Park. Zari envisions something similar that would connect east Portland's natural open spaces, including Johnson Creek and the Springwater Corridor, to make a several-thousand acre park on the other side of the Willamette River. Already several large parcels of undeveloped land on hills on both sides of Johnson Creek between Powell Butte and Leach Botanical Garden have been purchased with Metro's natural area bonds. The City of Portland's Bureau of Environmental Services has bought land along Johnson Creek for flood storage.

These sites are all largely contiguous, Zari continues. Combined, they would create an opportunity for a wonderful wild and natural park for east Portland, with additional properties still available for purchase to extend the publicly owned land. It is a vision, says Zari, that she hopes may someday come to fruition.

Zari's next dream—creating an East Bank Esplanade for Portland—was more difficult to accomplish than protecting Powell Butte. For over fifteen years, she tried convincing city officials of the value of developing a trail on the east bank of the Willamette River and making it available to the public. The challenges, she knew, were great. An interstate highway directly abutted the riverbank. There was a lack of land available for such an endeavor.

Undaunted, Zari began her advocacy with the commissioner of parks at the time, Charles Jordan. "I asked Commissioner Jordan, 'May I at least develop a plan to show what we could do? I will then take it to neighborhoods and to the planning commission.'"

It took some persuasion, but Zari was persistent, and Charles Jordan consented. The idea did not congeal all at once, however. It faced repeated setbacks. When Charles Jordan left Portland to accept a new job elsewhere, Zari next took the proposal to the new commissioner of parks, Mike Lindberg, in 1986. Lindberg loved the idea and gave her the green light. She introduced the concept of an east bank trail to the Portland Development Commission (now known as Prosper Oregon), hoping they would embrace it as they were preparing their urban renewal plan.

The idea, however, stalled. Years continued ticking away. The turning point came when a bright new star, Vera Katz, became mayor of Portland in 1993.

Zari appealed to the new mayor. Before long, she had influenced her to support the esplanade proposition. And not only in a lukewarm manner. Rather, Mayor Katz fully embraced the plan and took it on as her own. "Vera made the Portland Development Commission come up with the money to develop it," says Zari, appreciatively, grateful that her city now has a wonderful riverside trail on the east side of the river. Completed in 2001, it is used by thousands of residents and visitors year-round.

In 2003, after working with the parks bureau for twenty-two years, Zari was appointed director of Portland Parks and Recreation. The new job only increased her determination to make changes she believed critical to the livability of Portland. In the position of overseeing all of the city's parks, she attempted something never before done in any other US city. She fashioned an entirely new and innovative design of parks and park management.

She called it City Nature.

The idea was straightforward and came from the philosophy inspired by former Oregon governor Tom McCall. Nature matters, whether one is in the city or in a rural area. But sometimes we forget to put into our planning things that don't have a voice.

"Politicians and administrators put the needs of humans, especially children, very high, versus the needs of animals and the environment. Of course human wants and desires are important, but you cannot forget the environment. Our thinking," says Zari, "needed changing."

With characteristic tenacity, Zari set to work again. In the big picture it was really all a matter of semantics, but words, as Zari knew, sometimes make a pivotal difference in terms of people's mind-set. At the time, the outlook of Portland Parks and Recreation was just that: recreation. Recreation opportunities were considered the priority; natural areas were thought of as the leftovers. While natural parks amassed the largest acreage in Portland, they were allocated a pitifully small staff and budget to manage the areas. Many in the bureau, as Zari saw, didn't view natural areas as something that was truly an essential part of parks and recreation.

"This was the belief of our bureau, of our own people!" she says, amazed. "That's why I was determined to develop a whole new unit, and give it a new and robust name."

The name she conceived, City Nature, suddenly spliced two simple words that together gave a powerful emphasis on nature and its value to the city. Zari then worked to make it all she envisioned. She reorganized the department's structure, creating individual zones throughout the city park system. Each zone consisted of developed parks—baseball fields, playgrounds, and soccer fields—but held other attributes, too: natural areas. She wanted citizens to learn to value that "recreation" was more than just organized sports and activities. It also included going into and exploring the nature surrounding them.

To promote her idea and make this goal more pronounced, both internally and with the public, Zari changed the bureau's funding process. The creation of City Nature gave developed parks and preservation of nature equal footing—in terms not only of funding but also decision making, visibility, and awareness.

"My motivation for protecting our city's natural areas is a combination of my upbringing and the fact that I personally derive much pleasure, both physical and spiritual, from spending time outdoors, in nature. Each time I hike in the woods, it makes me reflective and contemplative. In this day and age in our society, we don't often have time to do that."

Zari grows more serious. "The grandeur of our state, our landscape, is so beautiful. We are exceedingly fortunate to have some remnants of it in the city! Having the ability to get out into it, to take a bus or a ten minute drive to walk to the woods, to listen to the birds, and be quiet, is rejuvenating.

"I have done what I did for Portland because I believe that every human being deserves this," she continues. "Unfortunately, there are many unlucky people who don't have the means to get out and enjoy the grand landscape of nature in our region—the Columbia River Gorge, or Mount Hood. I think they should have a bit of a similar experience—here—at low cost. I have tried for that. Why? Because it's the right thing to do."

When asked where these values she holds close have come from, Zari answers quickly. They were imprinted on her by her father, a deeply spiritual and religious Muslim. He was prominent in their city, kind and never imposing. He was not wealthy, she explains. He believed one only needs enough to live a comfortable life; one does not need more than that. Her father spent his time and wealth giving back and helping others. He was Zari's role model and inspired her to live a life dedicated to civic leadership. His were the same principles she saw in Tom McCall and his Oregon—governorship that was visionary and unselfish.

"We need leaders who can inspire people to understand that, while we may not benefit from 'it'—whatever it is—personally, right now, the real benefit is going to be in the future, for our children."

Like others who have lived through Oregon's history of remarkable land-use decisions, however, Zari admits to being worried today. "I don't see that happening anymore. The selfless leaders who made the decisive actions to assure that the way of life Oregonians have and cherish is preserved are no longer apparent."

In the place of courageous leadership, Zari fears, is an expanding "me" attitude, a "me first" sense of privilege, at the exclusion of generosity.

Zari looks thoughtful. "I believe in a sense of equity. One of the principles that permeated our house when I grew up was the importance of fairness. This is not just a fairness to people, but fairness in our relation to the earth. To live our lives so that we are fair to the environment, we are fair to the wild animals, and to our neighbor. I have always tried to live by, 'How would I feel if somebody did this to me?' I try to think, 'Whatever action you take, think about the other people.' Because, they are not really *others*. They could be you. They could be your child. They could be your neighbors. They could be your grandchildren."

Oregon's clean air, clean water, the grandeur of its natural landscape, and the generosity, openness, and friendliness of its people are things that

Zari values and deeply hopes will remain for many generations. Yet her question still remains: "How do we make sure that one hundred years from now Oregon is the way it is today? The way that I have witnessed?"

That searching question, and the desire to protect what remains of the Oregon she loves, have been Zari's twin motivating factors in her life.

"Farm and forestland need our protection. We must not remove the urban growth boundaries," she states firmly. "When I consider Oregon—the largeness of this state, the beauty of its varied landscapes—I believe it influences people's character. People are more open here, more welcoming, more willing to share. When you live in an environment like Oregon, it creates a generous nature."

Zari remembers a hike she took to picturesque Sahalie Falls on the McKenzie River in the Oregon Cascade Mountains. "It was a short hike, with two beautiful waterfalls. The landscape was absolutely breathtaking. Gorgeous fir and hemlock trees . . . the groundcover is sensuous and velvety . . . there are huckleberries here and there. It's just like a composition—you look at it, and it's like seeing a beautiful painting. You can feel your body relaxing . . . loosening up. It is like a beautiful person or child—you want to just hug it or squeeze it. You can learn so much from the beauty of nature that you can then try to emulate in your own home, in your own park.

"I don't see the beauty of nature as just grass and a few trees and a couple of playgrounds, as beautiful and green as they are. Rather, it is that sense of harmony it gives you . . . its texture, variety, color, smell! It affects your senses. And it makes you happier! It makes you more relaxed! Beauty affects our nervous system and our mental system."

Zari smiles, her expression reflecting the love she has deeply felt for the two important places that have shaped her life. "And I tell you, when you are happy and relaxed, you are kinder to other people and to other things.

"So that is what I mean about beauty. Beauty doesn't necessarily mean you *are* a beautiful person. Beauty *makes you* as a person."

If you know you are right, act like you are right.
If someone, or a group, is breaking the law,
or acting in opposition to the law, stand up
to it. . . . Because when something is as
intensely beautiful as Oregon,
that's what makes it worth protecting.

———————————

Henry Richmond

14

A Man with 1,000 Friends

HENRY RICHMOND

Observing Oregon while traveling down the I-5 corridor south of Salem, an out-of-state visitor is surprised. It is not the verdant beauty of the fertile Willamette Valley that amazes her. Nor is it the view of the picturesque coastal forests and mountains to the west. Rather, her incredulousness results from something that is not in her line of vision—something attributable to the work of a passionate lawyer who believed in and defended Oregon's signature land-use law.

The traveler asks, "Where are the malls?"

∾

"It was an opportunity to do something good."

That attitude would run throughout Henry Richmond's life. It would be the motivation behind his devotion to preserving farmland and forests, his ceaseless fight to protect Oregon's visionary land-use laws and, in time, his crucial role in upholding their integrity.

Born in Washington State in 1943, for the first twelve years of his life, Henry lived in Walla Walla, where his father was an administrator at Bonneville Power Administration. It was here that he was first exposed to public affairs in action, while observing his father stand up to political dishonesty, refusing to partake in many of the games that were played. Henry recollects how his father wanted him to have a firsthand education about government, even to the point of taking him out of school whenever something significant was happening.

"I remember my father giving me a note to be excused from class when I was in third grade because he wanted to take me to Pendleton to hear Adlai Stevenson giving a talk at the airport. Years later, I saw Stevenson again at the Democratic Convention in 1960 in Los Angeles. He was escorting Mrs. Eleanor Roosevelt, with her hand laid on his arm, into the convention. Those were just a few of the interesting exposures to political things I had when I was growing up," says Henry. "Perhaps from that, I always knew I wanted to do something in public life."

Henry's first foray into politics occurred while attending the University of Oregon School of Law after graduating from the University of California, Berkeley, in 1966. He became intrigued with the news headlines, splashed almost daily, about Governor Tom McCall's battle against the sewage on Oregon's beaches and other ravages on the coast. There was talk about improving coastal planning. It captured Henry's attention and aroused his growing sense of public service.

Putting his concerns into action, in 1970 he helped organize a group of students and created a consumer group. Its goal was to stand up to special interests if they threatened public health and safety. In 1971, the year he received his law degree, he drafted the articles of incorporation for Oregon State Public Interest Research Group (OSPIRG). The following year, only one year out of law school, Henry became the organization's staff attorney.

He soon met his first challenge.

In 1973, the pristine Metolius River in Central Oregon was threatened. The Metolius, a tributary of the Deschutes River near the city of Sisters, is long revered for its extraordinary beauty. Cold, crystalline waters and rich meadows are framed by towering Black Butte. The river is home to rainbow trout, bull trout, and mountain whitefish. River otters and beaver live next to the flowing stream. The Metolius is nationally renowned for its outstanding fly-fishing and is one of the largest spring-fed rivers in the United States. A developer, seizing what he thought was a tremendous opportunity, wished to develop the entire panorama. He proposed a 140-acre subdivision of the scenic Metolius Meadows.

Shrewdly circumventing land-use laws, he had not called his proposal a "subdivision," which would not be allowed under current zoning. Rather, he titled it "seven, twenty-acre plots." With this turn of phrase, the entrepreneur had evaded the prohibition and bypassed subdivision regulations.

The likelihood of preserving the unspoiled naturalness of the iconic setting seemed grim when Henry became involved.

In his role as the staff attorney for OSPIRG, Henry took action. He contacted the director of the Oregon Real Estate Agency. With a wry smile, he remembers the conversation as if it were yesterday.

"The director told me, 'This is in compliance of the law,' and then he boldly added, 'and there's nothing you can do about it!'"

Henry next wrote a letter to the director contradicting him, referring to the action as "misfeasance"—a performance of an act which might lawfully be done, but in an improper manner. Before the letter went out, though, Henry thought to ask his father's advice.

His father read the letter, then handed it back to Henry. He shook his head and looked at him thoughtfully. "'Why are you calling it misfeasance?' he asked me. 'It looks like *malfeasance* to me.'" His next words would profoundly influence Henry's thinking. "My father said, 'If you know you are right, act like you are right. If someone, or a group, is breaking the law, or acting in opposition to the law, stand up to it.'"

Henry was well acquainted with the term. *Malfeasance*: the commission of an act that is *unequivocally illegal* or *completely wrongful*. The word was unambiguous and far stronger.

Henry changed it in his letter.

The following week, Henry received a call from the Oregon attorney general's office. He was informed that Attorney General Lee Johnson agreed with Henry's appraisal. The developer *was* breaking the law. Outsmarted, the developer quit. The real estate director was later replaced.

The entire experience taught Henry a lesson that he carried with him throughout his long years of advocacy. "I learned not to be afraid to challenge people," says Henry. "Because when something is as intensely beautiful as Oregon, that's what makes it worth protecting."

Another issue came before Henry while he was at OSPIRG that deeply impacted him. This one had to do with wilderness laws. A California mining company had its sights trained on Oregon's third tallest mountain, picturesque South Sister, in the Oregon's Cascades. South Sister, at 10,358 feet, lay within the Three Sisters Wilderness area and was a popular hiking and climbing destination, with abundant glaciers, lakes, and meadows. The mining company hoped to snare a mining claim in the wilderness on the mountain's flank at a place known as Rock Mesa.

The landscape was a fifteen-hundred-acre lava flow, and the company hoped to excavate it to mine pumice. It insisted that the claim was legal, as it could be classified as a "commercially viable mine." South Sister was a deeply loved mountain and Henry was troubled. The visual integrity and beauty of the mountain would be forever marred if the company had its way. So, he came up with an idea.

Henry thought: Let's look at this problem in a technical way. Let's get the best geologist of the area, Dr. Ed Taylor, to research the claim for its viability. Let's go camping!

Henry, Ed, and a high school student named Peter Hayes took off to explore Rock Mesa on a three-day backpacking trip. Ed mapped the area and loaded heavy gunnysacks full of pumice, which Peter lugged back to the lab to be examined. What the geologist discovered was a problem with the material. The pumice stone was full of flecks of feldspar.

The use of the pumice, stated the mining company, would be for backyard barbeque grilling. The feldspar, though, made that highly impractical. As Ed noted in his report to the US Forest Service, feldspar destroys the cures of a grill. Therefore, Henry argued, this should never be classified as a commercially viable mine.

The USFS denied the permit.

"Every time now I see the beautiful South Sister, I feel pleasure knowing it is still intact," Henry says.

The next chapter in Henry's immersion into land-use planning began in 1974, nine months after Governor Tom McCall signed Senate Bill 100 into law on May 29, 1973. Senate Bill 100 was inspired and revolutionary, and was the first of its kind in the country. It came with the mandate that all cities and counties in Oregon must have comprehensive land-use plans over the entirety of their jurisdictions. Senate Bill 100's scope, strength, and depth far exceeded an earlier, less stringent version, Senate Bill 10, that passed in 1969. Senate Bill 10 had called for zoning and set deadlines, but it was weak.

In reality, in the early 1970s, most places in Oregon were not even zoned. Deadlines were routinely ignored. When forced to zone by state law, many jurisdictions began zoning for subdivisions, much in the same way that Washington and California and the whole country was doing—and continue to do. It meant that planning went from no zoning to zoning for subdivisions in half of all the cases.

"There always has been, and always will be, a very active real-estate lobby that's looking for income from transactions, not from asset appreciation," Henry explains.

But Senate Bill 100 changed that—at least in theory. Zoning was no longer an option; cities and counties were required to zone. What was needed, however, was some kind of administrative capability at the state level to ensure that the state land-use policies were implemented and not just sitting on the statue books. The Land Conservation and Development Commission (LCDC) was thereby formed—a land-use agency tasked to adopt the goals set forth in Senate Bill 100.

LCDC consists of seven unpaid volunteers who are appointed by the governor and confirmed by the state senate. Members are appointed to four-year terms and must represent certain regions of the state. Their role is to provide oversight, insure compliance, and coordinate state and local planning. LCDC guides the Oregon Department of Land Conservation and Development (DLCD) in implementing state land-use goals, provides technical assistance, collects data, facilitates studies, and participates in the judicial process.

Senate Bill 100's visionary formation of LCDC, in association with DLCD, became the force behind the face of Oregon. The bill established Oregon as the nationwide leader in land-use planning. It set specific goals, nineteen in all. Every city must have an urban growth boundary (UGB) to reign in urban sprawl. Land inside the boundary is specified to be used for a city or town's development. Land outside the boundary is to be preserved for agriculture, forestry, and maintaining a rural, natural state. The urban growth boundary was designed to be a highly innovative way to curb unplanned urban growth and to protect surrounding farm and forest land.

Yet a question remained with the enactment of Senate Bill 100. What was the best way to zone Oregon's premier agricultural lands? Every farmer knew that some agricultural land had greater value than others. Particularly, the Willamette Valley in western Oregon is extraordinarily fertile and among the most productive farm land in the nation. It is also located along the I-5 corridor, where cities and communities continue to rapidly expand.

The answer to the question of how to handle these important lands came largely from a farsighted idea, conceived in 1974 by a highly respected farmer from Zena, Oregon, Jim Smart.

Jim knew there needed to be specifics in protecting farmland or it could just be whittled away. He believed that designating farm potential by soil class was a powerful and clear tool. Every county in Oregon has a soil survey. But what legislators had recently concluded was that setting aside soil classed as 1 and 2 and even some 3s would provide adequate protection for prime farm ground.

Jim Smart disagreed. It was not enough. He met with Henry while he was still at OSPIRG to help Henry understand his reasoning. Henry recollects the conversation with a grin of appreciation.

"Jim had obviously carefully thought the question through. 'Look, Henry,' he said to me, 'What is currently being proposed to protect farmland is too narrow in scope. You need to adopt something stronger. When I drive my tractor over my farm, I'm driving over the 1s and 2s and 3s and I'm going over the 4s. If you don't protect the 4s, you don't protect the 1s and 2s and 3s.'"

Smart's idea became the foundation of Senate Bill 100.

Governor McCall appointed Jim Smart to the Land Conservation and Development Commission, aware that, within the agricultural community, all the famers knew and trusted Jim. He was a Republican. He was active in the Farm Bureau. He was respected for his total integrity in everything he did. As Henry relates, "Jim was clear. Soil classes 1 thru 4 should be set aside as agricultural land, and should be zoned Exclusive Farm Use."

While that standard helped save thousands of acres of farmland in Oregon, the enactment of the standards of Senate Bill 100 did not come without intense opposition. Many jurisdictions did not like the idea that they had to zone; even more, that there were specifications they had to zone according to. In essence, Jim Smart's farm wisdom, backed by the governor, meant that all the counties in Oregon with soil classes 1 thru 4—which included much of Yamhill, Marin, Polk, and Lane Counties—had to change much of their zoning from subdivisions to Exclusive Farm Use to follow state law.

Many balked.

As soon as the law was realized, dissention began arising and attempts were made to navigate around it. Counties squirmed and landowners griped. "Well, what the law really means is this," counties would appeal. "The law doesn't apply to building permits. It doesn't apply to comprehensive plan changes. It doesn't apply to annexations."

Henry saw this coming. He heard the arguments spewing forth from officials: "These goals don't go into effect for another year, you know." The mad rush was on to swiftly subdivide, and to find lawyers ready and willing to avoid and defy, citing "differences in interpretation."

"The handwriting was on the wall," Henry says. "Attempts were already in motion to deviate from the legislative intent of Senate Bill 100, thereby degrading the law."

Henry realized that there needed to be an organization with attorneys who could represent people in precedential cases—cases where there hadn't been a ruling on this point yet. There needed to be predictability of how the law would be administered . . . how the law *must* be administered.

In Henry's mind, Senate Bill 100 was worth fighting for. The law was perfectly clear: every city in Oregon was required to have an urban growth boundary; that meant, all 240 cities had to adopt them. And wherever there were Soil Types 1 thru 4 outside the urban growth boundary, the land was to be zoned farmland. Period.

In August 1974, Henry proposed to Governor McCall the idea of an organization that would provide free legal assistance to people who sought to present to judges and impartial judicial tribunals important questions of what the meaning of Senate Bill 100 was. McCall jumped on the idea. "This is great," Henry remembers him saying. "This is good. We need to raise some money. Why don't we raise $30,000 and I'll announce it."

That was all Henry needed to get started. Later that month, he composed a letter, sending it to people throughout Oregon. "We need to get 1,000 people to give us $100. That's our goal but Governor McCall will announce it when we get 300. Will you be one of the first members?"

Thus, 1000 Friend of Oregon was born.

State representative Betsy Johnson was the first member, with a $100 contribution. "Sounds good to me!" she said, getting out her checkbook and writing a check. Before long, more people began pledging and giving money. The envelopes started piling up. Henry formed an advisory board. Before the year's end, he had raised $30,000. On January 8, 1975, Henry drove to Salem to talk to Governor McCall.

Henry assumed the meeting would consist of writing a news release that McCall would later give to someone to take down to the press room to share. The governor, however, surprised him.

"Come on, let's go!" Henry remembers him saying. "*Come on, Hen!*" McCall strode down to the press room, and Henry followed. Marching in, the governor was beaming, full of confidence. He gave a rousing speech, announcing the formation of 1000 Friends of Oregon. He stated many illustrious names, listed the advisory board members, and applauded Henry Richmond for his great vision and courage to defend what was special about Oregon.

And then the real work began. After forming 1000 Friends, Henry left OSPIRG to launch the new organization. As he had anticipated, challenges to Senate Bill 100 arose almost overnight. The first case was in Marion County, which was conducting an area-wide re-zoning of twenty-four thousand acres outside the urban growth boundary. It was all agricultural land. County commissioners balked at zoning it for exclusive farm use.

"That law doesn't go into effect until a year from now, and even if it did, we don't think it means what you say," Henry remembers them stating. "And we're going to zone it for ten-acre subdivisions."

Henry explained that was illegal. They didn't care and went forward with their plans. Henry appealed.

1000 Friends of Oregon v. Marion County quickly became one of the state's biggest cases and tests of Senate Bill 100. Robert Stacey, a 1000 Friend's attorney, handled the case. The law was upheld. The courts found that Marion County had to comply with the law and zone according to the law's standards.

Other challenges to Senate Bill 100 quickly followed. One of the most serious threats testing the strength and integrity of the law Henry took all the way to the Supreme Court of Oregon. It had to do, of all things, with utility corridors. The case was outwardly simple: "Did utility corridors have to follow the strict guidelines, or did they not?" and "Do the goals of Senate Bill 100 apply to annexations?" The City of Klamath Falls, hoping to circumvent the law, concluded they did not. Rather, the city reasoned, utility corridors could be cleverly used to bring in land outside the urban growth boundary for annexation.

A developer had conjured up the ingenious strategy this way: There was land he wanted to subdivide. It lay, however, twelve miles out of town, outside of the boundary, therefore was subject to falling under the exclusive farm use specifications. The developer argued that it was *not* outside the line, that it was indeed *attached* to the inner city, by virtue of a narrow snake of land: a utility corridor.

If the utility corridor was not subject to being either in or out of the boundary, it could act as a conduit—connecting a 120-acre parcel ripe for subdivisions to the city itself. Following this argument, the City of Klamath Falls eagerly annexed the property.

Henry, however, realized the dangerous precedent that was being set. He sued. A Klamath Falls resident, Paulann Petersen, who later became Oregon Poet Laureate, was the plaintiff.

They lost the case.

Henry appealed to the Oregon Court of Appeals: "*Peterson v. the City of Klamath Falls.*"

They lost again.

Henry did not give up. He appealed to the Oregon Supreme Court. It was the first case to go to the Oregon Supreme Court for the defense of Senate Bill 100. And at the highest court of Oregon, Henry won the case. The court reversed the ruling of the lower courts and the court of appeals. The Oregon Supreme Court ruled that the goals of Senate Bill 100 did apply to annexation decisions. The ruling was unanimous. Henry says he will never forget the opinion arriving in the mail to the 1000 Friends office.

"It made my day!" he recounts animatedly. "More than that, it made my life! It gave 1000 Friends and our cause a lot of credibility. It gave Senate Bill 100 the strength it needed to make a difference. It was like saying, 'This land-use planning idea? It is not going to go away!'"

And it did not go away. In its first three years, Henry's organization won 90 percent of its cases—an unprecedented beginning. Senate Bill 100 became the foundational law it was intended to be. Forty years after the enactment of Senate Bill 100, over 15.5 million acres of Oregon's rural lands are zoned exclusive farm use, protecting them from patchwork development, being swallowed into subdivisions, or subsumed by urban sprawl. As well, approximately 9 million acres are protected and zoned for forest uses to maintain a forestland base, as overseen in DLCD's Forestland Protection Program.

Today, 1000 Friends of Oregon still actively pursues the mission to guard the integrity of Senate Bill 100. It strives to maintain family farms and forests, to safeguard Oregon's livability, and to conserve natural areas. Its focus remains the same as when Henry founded it—to defend Oregon's quality of life.

With an expression of deep satisfaction, Henry, now seventy-three years old, looks out over Star Mooring Farm on the Willamette River where he lives with his wife of thirty-seven years, Ruth. The land was her parent's home, where they had raised purebred Black Angus and tended hazelnut orchards. After her parents passed away, he and Ruth moved back to the farm in 2011, converting the cattle-breeding operation to hay pastures, hazelnut groves, and grapes. He says that without Oregon's land-use laws, all of the farms up and down the green valley would be gone, and subdivisions in their place. That thought only makes him appreciate the beauty and value of the rural land all the more.

"I would not want to see the laws that protect Oregon's beauty be degraded," he reflects thoughtfully. "I regard the land as a gift of God. You are a trustee and the land is the trust. You and I are not going to live forever. After we are gone, somebody else has to be the trustee. I hope they will be there."

He takes in the sweeping landscape surrounding the farm. "Oregon is so loveable because it is so beautiful. And in some way, I associate that beauty of the land with all the wonderful people with whom I've had the privilege over the years to work with. The best part, the lucky part, has been meeting all these people all across the state of Oregon, who are drawn, and continue to be drawn, to this terrific cause."

Henry Richmond indeed is a lucky man. He has a happy marriage. He has a close family. He has saved a large portion of Oregon for all of us to enjoy. He has stayed true to his ideals, through periods of disappointment and through periods of success, for a lifetime.

Even more, he is a man with 1,000 friends.

I have traveled over three million miles.
I could live in any place in the world.
I want to live here, in Oregon.... We need
to protect Oregon, both visually and physically.
Because the underlying thing?
It is the quality of life.

———————

John Gray

15

The Shortest Line
JOHN GRAY

A love of Oregon and the values instilled during his childhood would inform his life philosophy: the importance of good planning, defending strong land-use laws, and providing opportunities for people to live in harmony with their environment. In time, they would also inspire him to create a multi-million-dollar legacy initiative to support these ideals through a unique vehicle: land trusts.

༄

If the engineering or agriculture departments at Oregon State College had not had long registration lines in 1936, the face of Oregon might be different today. There may not have been a Salishan Resort on the coast, a Sunriver near Bend, or a Skamania Lodge in the Columbia River Gorge. Even more, the achievement of Oregon's land-use planning laws may have been considerably more tenuous.

But there *was* a short line, as John Gray contemplated which major to select for his degree. "That's why I chose it," says John, matter-of-factly. "It just seemed logical. Kids don't know what they want to do. I didn't either. Business had the shortest line. I registered and just got on with it."

"Getting on with it"—with confidence, decisiveness, and lots of hard work—became a hallmark of Gray's life. Part of that came from his experiences growing up, which instilled in him a strong work ethic. But a compassion for others and an understanding of what it meant to overcome hardship, always would drive his decisions too, as well as a true affinity and appreciation for the rural landscape in which he grew up.

"The farm was a big influence in my life. Just seeing good land."

The land, and the good people who farmed it, played a large role in his childhood. He admits if he hadn't grown up the way he did he might have been a different person. Never one to feel sorry for himself, he is objective about the conditions his family faced.

"We were very, very poor when I grew up. I was born in Ontario, Oregon, in 1919 and we lived on a farm. My mother had been a schoolteacher before she got married; my father was a rancher. Things changed when I was six; my dad died."

How she was going to raise alone three small boys—John and two younger brothers—became his mother's focus. She returned to Corvallis, where her parents lived, and took a teaching job at a one-room country school north of Monroe. It was a farming community, and they moved into a tiny home on a large farm.

"We were tenants and leased the little home for $2.50 a month. We lived there for eleven years; that's where I was raised. My brothers and I all went to my mother's school. All the grades were together; it was a one room school, and during the Depression. At that time, a school teacher was really looked up to."

The farming families in Monroe were very friendly and took John and his brother under their wings. "A lot of the dads were our substitute dad," he remembers appreciatively, and they held similar values to his mother's. "What she wanted," John says, "was to see us all get through college. We all did. A lot of our ethic came from just being around good people."

John never forgot the connection between 'being around good people' and making good things happen. But he also knew that making good things happen required lots of work and study. His mother made sure that education was an important value in their family.

"I was a good student; I studied hard," John says. "I got a scholarship to Oregon State College. Unbeknownst to me, the principal of Monroe High School applied for me. He got me a four year, tuition-free scholarship. At the time it was $100 a year. I worked too for all of the farmers. I also drove the school bus for Corvallis High School when in college. I had to work. I needed the money to go to school."

That single-minded determination to carve out a future for himself—while at the time not knowing exactly what that might be—carried over to other things: his personal life. Through college, with his nose to the grindstone, he had never dated. That changed when he took a ballroom dancing class. His brothers, who were also attending Oregon State College, encouraged John to do it. He saw Betty there for the first time, when she was a freshman and he a senior.

"That was it," John relates. "She was the only girl I ever dated. And all we had were a few milkshakes together at Oregon State because I was soon called into the Army to active duty."

John and his two brothers all served in World War II. John was sent to Europe with the 101 Airborne Division and later the 82 Airborne Division, earning the rank of lieutenant colonel and a Bronze Star Medal. During that time, he corresponded with Betty from afar. After five and a half years, he returned home from overseas. Betty was there to meet him at Fort Lewis in Washington State. She drove him to McMinnville to see her family.

They married four days later.

"We had both waited for each other," he explains. And for the next fifty-seven years, they made up for their initial separation by supporting each other and raising five children together.

John met each opportunity that arose in his life with the same focus and energy. "I've always tried to live by the adages 'go the extra mile' and 'do more than expected,'" he relates. John applied to Harvard Business School, was accepted, and earned an MBA in 1947. Upon graduation, he and Betty returned to Oregon, where he began work for a forestry equipment company. He was then offered a job at Oregon Saw Chain, which became OMARK Industries.

Demonstrating his characteristic tenacity, he quickly rose through the ranks. Before long, he became general manager. Five years later, he bought the company.

For the next thirty years, John was the president and chairman of OMARK Industries, known for its manufacturing of chain for chain saws and other logging equipment material. With numerous overseas operations, and plants scattered in many different countries, John traveled extensively. Through witnessing many parts of the world, he came to a strong conviction.

"I have traveled over three million miles. I could live in any place in the world. I want to live here, in Oregon. I like the variety of Oregon's landforms and the varied climates we have. I like the fact I can move around the state and see different things. And I like the quality of its people."

In the 1960s and 1970s, John became a strong supporter of statewide planning and zoning. Oregon was rapidly gaining population; the fastest growing region was also the most fertile—the Willamette Valley. With urban sprawl and suburbanization threatening the state's farm and forestlands with rampant, unplanned development, the passage of Senate Bill 100 was paramount in reining in poorly designed, scattered growth. SB 100 created a defined structure for thoughtful, comprehensive zoning for every town and city in the state of Oregon. It made communities responsible for preserving irreplaceable scenery, natural resources, farmland, and timber lands.

John became heavily involved in the battle to protect the goals of land-use laws, which were at times threatened by noncompliance or even repeal. In 1975, two years after the passage of SB 100, he became one of the first board members of 1000 Friends of Oregon—an organization dedicated to protecting prime farm and forestland from development. The following year, when SB 100 faced being abolished by an initiative on the Oregon ballot, Measure 10, John was on the steering committee of Citizens to Save Oregon's Land, a group opposing the measure. Measure 10 also sought to put an end to the Land Conservation and Development Commission (LCDC), established with the passage of SB 100 as the oversight committee tasked to oversee compliance of local planning with statewide goals.

John's personal philosophy helped to frame the opposition's stance. As a father who routinely took his family—Betty and their five children—backpacking on five- to ten-day camping trips, he believed that wild land needed to be saved for future generations of Oregonians. He worked closely with land-use champions Henry Richmond, Hector Macpherson, Larry Williams, and others, and with the fierce support of Governor Bob Straub, to define the dispute clearly. "The issue boiled down to those who said that LCDC was going to take away your property versus those who said we are trying to protect Oregon for our children and grandchildren."

Oregon voters listened. Once more they spoke out to defend their land and retain planning. Measure 10 failed.

"My love has always been for good planning," says John. "It is only good, common sense. I like to see things well-developed. I do not like unplanned growth." And whenever the issue arose, John was there to speak out for protecting Oregon's quality of life. His desire for intelligent zoning and planning resulted in three things for which he is perhaps most well remembered: Salishan Lodge, Sunriver Resort, and Skamania Lodge—located respectively on the Oregon Coast, Central Oregon, and the Columbia River Gorge.

All three developments are among Oregon's most prominent resorts, sensitively designed and in keeping with the local natural environments. John created them in spots to which he had deep personal attachments. His goal was for thousands of people each year to experience Oregon in the way that he intended—respectfully and with appreciation. Along with John's Landing—a residential and commercial district he redeveloped along southwest Portland's industrial riverfront, named for B. P. John's Furniture Company—these developments are his most visible achievements. Yet those things he has done behind the scenes are perhaps even more lasting and significant.

In a quiet, modest, unpretentious manner, John is a generous philanthropist and vigorous advocate for the environment, health care, affordable housing, and protection of Oregon's resources. He also is a strong supporter of environmental education. John has been a staunch proponent of outdoor learning and a long-standing Oregon program known as Outdoor School.

The idea of Outdoor School originated in the 1950s in the cities of Medford and Prineville. Its scope grew substantially when the program became established in Multnomah County in 1966. The goal was groundbreaking and far-reaching: to teach all sixth-grade students in the county and surrounding region about native plants, animals, soils, and water during a week-long, school-sponsored program in an outdoor setting. Students would learn hands-on, field-based science directly, while living at different campground sites in the region.

Today, Outdoor School reaches students across the state and is recognized as "one of the longest-running and most successful environmental education programs in the nation." In 2016, Oregon became the first state in the country to fund a full week of environmental education, with a bill signed into law providing $24 million in funding to support fifth- or sixth-grade students to attend Outdoor School. The Gray Family Foundation, motivated by personal priorities set by John, was instrumental in persuading Oregon voters to support the measure by a margin of more than two to one.

Another innovative endeavor created by John might have the greatest impact on the state he loves. "From an early age, I knew that there was something very special about Oregon and felt a strong loyalty to my homeland," John professes. "Later in life I recognized that if I wanted my great grandchildren to have this same connection to the land, I needed to be active in protecting this incredible place before it is too late." In 2011, at age ninety-one, John designed and endowed a first-of-its-kind, ten-year, ten million dollar program to strengthen Oregon land trusts.

Land trusts are nonprofit organizations formed with the mission to protect, preserve, and care for important landscapes and natural areas. In Oregon, twenty-one land trusts, from the Oregon Coast to the Wallowa Mountains, are members of the Coalition of Oregon Land Trusts (COLT). Defending thousands of acres, they work to ensure special lands are preserved for future generations. Working with willing landowners and community partners, land trusts use strategies such as conservation easements or fee-title acquisitions to accomplish this objective.

Always supportive of the idea of forging public/private partnerships, John is a firm believer in the land trust concept to permanently protect important areas—magnificent open spaces, rich agricultural lands, rivers and wildlife habitat. His endowment, the Gray Family Land Trust Initiative, is a bold family enterprise that immeasurably benefits irreplaceable natural areas throughout Oregon. Its overarching purpose is to provide opportunities for nonprofit land trusts to respond to the needs of local Oregon communities.

"This gift is meant to be a catalyst," John says. "I want Oregonians to take an active role in being stewards of this land, and to join my family in committing to funding conservation and restoration. Our gift is only a slice of the pie. We need more people to take a stand for Oregon."

The initiative continues to afford critical funding for numerous conservation projects. It provides essential aid to the national Land Trust Alliance and the Coalition of Oregon Land Trusts. As with all John's undertakings, his donation reveals profound concerns: the welfare of future generations, leaving things better for them, and caring for Oregon.

"My love for Oregon has come from what I have experienced in my life: the farms, the mountains, the hiking, the beach. We need to protect Oregon for its air and climate control," he says, stating his concern over climate change and the devastating effects it poses to the globe. Healthy Oregon forests, he attests, play an important role in carbon sequestration.

"Oregon is a wonderful place people can enjoy—both visually and physically. Because the underlying thing? It is the quality of life."

John pauses, then with conciseness that comes from a deep wellspring in his heart, he adds, simply, "I just like to do good things on the land."

For John Gray, it began with a short line. For the rest of us, what resulted is an enduring legacy for Oregon longer than a lifetime.

I hope there will always be Oregonians
who will step up for wild places . . . for old-growth
forests, natural open spaces, and clean, free-
flowing water. We must do it for the sake of our
children, for all young people, and for the benefit
of future generations of Oregonians. Because
at the end of the day, everything's coming from
the earth. It is fundamental to our existence,
not only to our quality of life, but to our survival.

───────────────

Charles Ciecko

16

Serendipity

CHARLIE CIECKO

It is October, and the salmon are spawning once more in the wild and scenic Sandy River at Oxbow Regional Park. Hundreds of people will be gathering to watch, thanks to the dedication of a man who devoted much of his life to safeguarding the prized river from many threats and to helping others learn to appreciate its value. Today children and families will be coming to the well-loved annual event he developed—a celebration of the return of Chinook salmon, where volunteers, naturalists, and tribal representatives will teach of the salmon's critical importance to Oregon's ecology and tremendous influence on human quality of life.

Growing up in the 1950s in a small community surrounded by farms, forty miles from downtown Chicago, Charlie Ciecko knew freedoms few children today experience.

"There were four kids in my family. We lived in a modest home. My mother would tell us, 'You can't be in here; it's too crowded, and I've work to do!' But she would say it with a laugh, as she scooted us out the door," Charlie recollects, with a slight grin. "During the summer we would leave in the morning, come home at lunch, and be shooed out again. We would explore the lakes and streams and marshes close by. We'd come home at dinner, and be out again until dark. There was never a worry. Nobody had to worry about their kids."

For a young boy like Charlie, it was heaven. It meant he could wander unfettered and free in the rich natural world around him. He was not alone. It was the way many children across the nation grew up just a generation or two ago.

"It's a different world now," adds Charlie, with a tone of both regret and concern. "Parents wouldn't dare to let their kids just wander."

Charlie worries because those wanderings did something more than just provide a venue for enjoyment. They also instilled an intrinsic connection to outdoor life and the natural world. Today, he feels, that attachment seems to be greatly diminished, resulting in part from the disappearance of direct lines between kids and nature.

Charlie acknowledges that his parents, as well as the close contact with nature just outside his door, helped form his appreciation for the natural world. "My father encouraged me to be involved with everything outdoors. Every family vacation was a camping vacation. If it involved boating or fishing, I was hooked. It made me a water guy. I love being on or near water."

Beginning college at the University of Northern Colorado, Charlie studied science for two years. During that time, he met Terry, who would become his wife. In 1973 they moved to Corvallis, Oregon, sight unseen, where Charlie enrolled at Oregon State University to pursue a wildlife biology degree. "But the more I reflected on it, the more I came to realize that I liked the notion of having the people part mixed in with the natural resources." Ultimately, he ended up in the School of Forestry and earned a BS degree in resource and recreation management. After graduation in 1975, he began looking for a job.

Serendipity struck.

A friend of Charlie's, whose father worked for Multnomah County Parks, mentioned that he'd heard about an opening at one of the parks east of Gresham, Oregon. Charlie made an appointment with the director. When they met, the director told him that the supervisor at Oxbow Park had recently taken ill. While not offering Charlie a job exactly, he said there was an old house at Oxbow Park that was vacant.

"We need someone to live in the house and help with park maintenance right now. It's a temporary job," Charlie remembers the director saying. Charlie was willing to do that. He asked, though, if there was a chance for him to apply for the permanent job that the prior supervisor had just vacated. Thinking for a moment, the director replied, "Sure, why not?" and left it at that.

Six months later, Charlie ended up getting that job as Oxbow Park supervisor. He was twenty-three years old. "People would come to the door at the house asking if my dad was home," says Charlie with a laugh.

The job could not have been a more perfect fit. He resided in the park with Terry and their two sons. The thousand-acre natural area was on the picturesque Sandy River. There was the water he loved for boating, fishing, and rafting. There was abundant salmon, steelhead, and wildlife. Fifteen miles of trails wended through the river canyon, a preserved remnant of ancient forest.

The opportunity was also timely. During the same period he was beginning his career, he began observing, with increasing dismay, the push by federal agencies to liquidate the last of Oregon's old-growth forests. He perceived something else, as well—the strength of public opinion.

"The real opportunity for overcoming some of the myriad threats to Oregon's natural resources exists through public engagement," Charlie notes. "I saw that if you could help promote a conservation ethic in the urban areas—saving nature in the city—you would have a better chance of succeeding in the rural areas as well."

Supporting and developing that ethic became Charlie's lifetime goal.

While at Oxbow Park, Charlie embarked on several grassroots activities above and beyond the daily responsibilities of his job. He researched, wrote, and illustrated an educational book, *A Teaching Guide to Oxbow Park*. He hoped it could be made available to teachers, scout leaders, church group leaders, and others to provide an educational component to their park visits. The Oregon Parks Foundation awarded him a grant to fund its publication, and Charlie's book was immediately successful. It also formed the beginning of a robust environmental education program that today is still ongoing at Oxbow Park.

Charlie's environmental efforts didn't stop there. After witnessing a wide variety of illegal fishing activities related to spawning fall Chinook salmon occurring in the park, he endeavored to increase public awareness. While pressing the Oregon Department of Fish and Wildlife to close fall Chinook spawning areas within Oxbow Park to fishing during the spawning season, Charlie organized an event called the Salmon Festival at Oxbow Park.

The first year, the Salmon Festival consisted of a few guided river-and-fish-viewing walks. Interest, however, quickly grew, and word spread. More and more people wished to participate and to learn about salmon. The Salmon Festival soon became a major annual event involving many partners and a variety of salmon related educational, entertainment, and cultural activities.

Today the festival is immensely popular and attended by hundreds of visitors each year. What is meaningful to Charlie is that, over its twenty-year run, tens of thousands of people have learned about salmon. As well, resulting from his continued advocacy, the Oregon Department of Fish and Wildlife ultimately closed the spawning areas to angling during the spawning season and became a partner of the Salmon Festival.

Charlie's career continued to ascend. In 1984, after nine years at Oxbow, he was appointed Director of Multnomah County Parks. It was a position he enjoyed for ten years, and allowed him to continue his role championing local environmental issues, especially efforts to protect the Sandy River that he loved.

Charlie effectively lobbied to have the lower Sandy River included in the Columbia Gorge National Scenic Area in 1986. He also strongly advocated to have portions of the Sandy River included in Senator Mark Hatfield's Omnibus Wild and Scenic Rivers Bill in 1988—an amendment to the Wild and Scenic Rivers Act of 1968. The new bill designated forty-one additional river segments in Oregon as scenic, wild, or recreational. Again, Charlie was successful. Portions of the Sandy River were declared wild and scenic, offering greater protections to their integrity.

Then, in 1993, taking him and others by surprise, Multnomah County made an astonishing announcement. Because of limited financial resources, the decision was made that the county would focus on health and human services. Multnomah County would get out of the parks business altogether. That year, in a major move, Multnomah County government transferred all responsibilities for parks to an entirely new agency, the Metropolitan Regional Government, or "Metro." It was a dramatic change for everyone involved.

Metro, established in 1978 by a statewide ballot measure, was designed as a regional governing agency and a metropolitan planning organization, with an elected president and six councilors. In the four decades following its formation, its presence and role in the Portland metropolitan region have increased substantially.

Metro's duties include the management of planned growth of three counties—Multnomah, Clackamas, and Washington—while serving twenty-five cities in the Portland metropolitan area. The agency is responsible for maintaining the urban growth boundary—the border between urban and rural lands, with the goal of reducing urban sprawl. The regional government also oversees the Oregon Zoo, Oregon Convention Center, the Portland Expo Center, Portland's Center for the Arts, metropolitan garbage transfer stations and recycling facilities, and the management of all Multnomah County parks.

With the transfer, Charlie lost his job as director of Multnomah County Parks. However, serendipity struck for him once again. While negotiating the conveyance of his duties to the new regional government, Charlie was approached by an executive at Metro, Rena Cusma. Cusma explained that Metro had recently adopted something called the Greenspaces Master Plan. The regional plan provided a framework to protect important natural areas through an innovative partnership with committed groups—governments, nonprofits, land trusts, citizens, businesses, and Metro. The goal of the plan was to preserve wildlife corridors, encourage environmental awareness so that citizens would become involved advocates, restore green spaces in neighborhoods where they were all but eliminated, and work to preserve significant watersheds, fish resources, and native plant and animal life in the growing urban environment.

Cusma knew of Charlie and his work. She told him there would be a new department starting up called Metro Regional Parks and Greenspaces. She offered Charlie the opportunity to become its director. Charlie's job, as Cusma explained it, would entail moving toward a regional system of parks throughout the three counties and creating a system of interconnected, open spaces for the multi-county region. Cusma also wanted Charlie to assemble for voter consideration a regional bond measure for acquiring new natural areas and stream corridors.

The scope of the new position, of course, was thoroughly concordant with Charlie's own personal goals. Beyond protecting critical open spaces and preserving wildlife habitat and water quality, it provided opportunities for people to experience nature close to where they live. In a way, the objective of the Greenspaces Master Plan was a return to the life that had grounded him in his youth and continued to be his inspiration.

Charlie accepted the assignment readily. From 1993 to 2003, he undertook the challenge to make the vision a reality and was immensely successful in shepherding the plan. In 1995, the $135.6-million Open Space, Parks and Streams bond measure was approved by voters. As a result, eight thousand acres of wetlands, meadows, and forests, and over fifty miles of stream and river frontage, were acquired from willing sellers over the next seven years.

During the same time, Charlie oversaw the creation of a volunteer program to help steward the new lands, and worked to expand nature-based education and recreation programs. He was extremely fortunate, he expresses with gratitude, to have a talented and dedicated staff working for him. "They deserve a lot of credit for the early success of the department."

Today, resulting from its solid foundation, Metro's Parks and Greenspaces Program continues to flourish. Successive bond measures have been passed, and by 2017, the number of protected spaces now surpasses seventeen thousand acres in and around Portland, including ninety miles of river and stream banks, numerous regional trails, and hundreds of community projects.

The accomplishments of the Greenspaces Initiatives have been deeply satisfying to Charlie, and since his retirement in 2010, after thirty-four years working with parks and open spaces, his passion for nature remains undimmed. Reflecting on his over three decades of conservation work, Charlie looks back upon his early years when it all started for him.

"In the 1950s and 1960s, it wasn't a wealthy time, but that generation stepped up. They were willing to pay more—for good schools, for infrastructure development. Maybe they didn't like it, and they surely complained about it, but they did it. I think we were a better country for that," Charlie expresses thoughtfully.

It encourages Charlie, though, that Oregonians *are* willing to pay for it—to secure those precious open spaces. "I hope there will always be Oregonians who will step up for wild places—for old growth forests, natural open spaces, and clean, free-flowing streams and rivers," he avers. "We must do it for the sake of our children, for all young people, and for the benefit of future generations of Oregonians."

Charlie understands that these resources are not merely favorable to have, but critical to the future of the earth that he loves. "It is one of our primary responsibilities to try to pass things on to our kids in a better condition. Because at the end of the day, everything's coming from the earth. It is fundamental to our existence, not only to our quality of life, but to our survival."

Charlie adds a final thought, one underscoring everything he has worked for his entire life. It is his greatest concern, as well as his hope.

"Today our natural resources and environment are under siege. It has never been more important for all citizens, especially our youth, to become informed and actively involved in our democracy."

Why is it important to protect Oregon's native landscape? It is because it is our home. It is our habitat; other species depend on us to survive. If those of us who live here don't protect it, it will not get done. We owe that to those who will come after us, just as we have benefited, or suffered from, what has been done in the past.

———————————————

J. Michael McCloskey

17

He Saved the Land; She Saved the Sea
J. MICHAEL McCLOSKEY

He spent his youth exploring Oregon's mountains and forests. Then, as a man, he left the state to fight for the passage of the Wilderness Act by Congress. In time, his continued advocacy helped enable the National Wilderness Preservation System to include 110 million acres of designated wilderness, with 2.6 million acres set aside in his beloved home state, Oregon.

∾

Those grateful for the adoption of the several momentous environmental bills and the creation of forty-seven official wilderness areas in Oregon can be glad that Michael McCloskey, after graduating from law school in 1961, changed his mind.

"Up to that time, I thought I wanted to be a politician," he says, looking back at his life choices. "I was born in Eugene, Oregon, went to Harvard for college, then University of Oregon for law school. I had been president of Young Democrats at college and law school. I ran for the legislature two years out of law school. I was nominated as a Democrat in Lane County and did reasonably well for my first effort."

He lost the race but learned something about himself. "I didn't like campaigning. I didn't want to talk about how the budget could be shrunk and taxes could be cut. What I wanted to talk about was conservation. I enjoyed hiking and climbing mountains and was cochairman of the conservation committee of the Obsidians—a climbing club in Eugene. I wanted to work for some nonprofit organization, for some cause."

Before long, doors opened. Michael found jobs with two conservation groups. Each wished to hire him to do fieldwork. Both organizations,
working together, paid half his way. One was the Federation of Western
Outdoor Clubs. Incorporated in 1932 and still active today, this union of
hiking and conservation organizations is mutually engaged in promoting
the protection of scenic lands, wilderness areas, wildlife, native plants, soils,
and waters in western states.

The other was the Sierra Club, at the time a grassroots organization of
sixteen thousand members and twenty-five paid employees, headquartered
in San Francisco, California. It hired Michael to be its first field representative in the northwest, based in Oregon. After working in this half-time
capacity for four years, the Sierra Club asked Michael to join the staff in
San Francisco. He accepted.

Michael stayed with the Sierra Club for the next thirty-five years.
In 1969, he became its national executive director and later chairman,
remaining in that role for over three decades. Today, the Sierra Club has
three million members and supporters and has become the nation's largest, longest-standing, and among the most effective conservation groups in
history. It all began for Michael, though—just twenty-seven years old and
recently out of law school—as a traveling adventure in a Volkswagen and
sometimes on horseback.

"The Federation and the Sierra Club had me going all around the
northwest—Oregon, Washington, Idaho, and western Montana—to make
contacts and give them advice how to organize conservation efforts more
effectively," Michael says. "As I went around, I heard stories about what was
working and what was not working."

Several things, Michael quickly learned, were not working in Oregon.

There was a threat to the Minam Valley in northeastern Oregon, which
was slated to be logged. The unspoiled Minam Valley, located in a remote
section of Wallowa County, known for its outstanding scenery, fisheries,
and wildlife, enfolds the remarkably pristine Minam River, now designated
a wild and scenic river. Logging was also planned for the Oregon Cascades'
picturesque Waldo Lake in the central Oregon Cascades, recognized as
one of the largest and deepest natural lakes in Oregon and one of the purest
lakes in the world.

As he deliberated about the best positive action in each case, Michael remembered three things he had learned in a class at Harvard while majoring in American government. He called them the three rules of lobbying.

"The first one," Michael explains, "is to know exactly what you want. You can't sell a fuzzy idea. The second is to identify who is the prime entity you need to persuade. It may be a court or legislative body or an administrative officer or even a private owner. Number three is the most important and frequently forgotten. You need to tailor-make your strategy to the specifics of each case. In some cases, it may be running a full-page newspaper ad. In others, it may be organizing around pieces of legislation. In another case it may be a big public relations campaign. And in others, it may require filing a lawsuit. You have to figure out what will work best."

Michael's thoughtful approach as a field representative set a tone that would define his forty-year career. He was, by all accounts, a good strategist. He was intelligent. He was always polite, nonthreatening, never abusive, and did not make far-out statements. He had credibility. His practical, down-to-earth style and integrity won him friends and, even when they did not agree, respect from those he worked with.

He was also innovative. In 1962, he worked to get the high mountain policy established—a strategy that stopped commercial logging at high elevations in the Oregon and Washington Cascades. The decree laid the groundwork for protecting the Minam Valley, which later became included in the Eagle Cap Wilderness, and for preserving the sparkling waters of Oregon's cherished Waldo Lake.

But it is when describing his role in the passage of the Wilderness Act in 1964—one of the most important pieces of environmental legislation ever enacted—that he smiles. Above all, it captures his resourcefulness.

Proposed in 1961 by Howard Zahniser of the Wilderness Society, the Wilderness Bill had been stuck, unmoving, for three years in the committee on interior and insular affairs. Its near death and the wrath of environmentalists and a large number of citizens were the result of one man, Congressman Wayne Aspinall, a lawyer and politician who represented western Colorado. Aspinall, chairman of the committee, was well known for his anti-environment stance.

As Michael explains, the congressman strongly believed that creating wilderness would negatively impact the miners, irrigators, and ranchers in his district. Having received little counterpressure from his constituents for three straight years, Aspinall aptly used one stalling tactic after another, making certain that the Wilderness Act would never see the light of day.

The Sierra Club's executive director at the time, David Brower, sent Michael to Colorado to see if there was anything he could figure out to do. "I started from scratch," Michael recollects, "contacting the conservation groups in Denver to see whether any of them had personal contacts on the western slope. They said no. So I flew to Grand Junction, rented a car, and had the idea to visit a sporting goods store. I asked the clerk, 'Would you know of anybody on the west slope who uses wilderness and would support efforts to protect it?' He said, 'Sure.' I said, 'Who are they?' He said, 'Packers and guides.' I asked him, 'Do you know how I might find them?' He said, 'I have a list right here from the Colorado state wildlife department.'"

That list, providentially, included names and addresses. The locations, Michael was glad to see, were not post-office boxes but real home or ranch addresses. Writing them down, he thanked the clerk and left—knowing exactly what he was going to do next.

Michael took off down the highway. One after another, he visited the residences or places of business, knocking on their doors. After introducing himself when they answered, he asked the guides and packers how they used the wilderness-type areas. He then asked whether they cared about keeping those areas intact for them and others.

"They said, 'You bet!' I next explained there was legislation pending that would help protect these areas for the future. I finished with, 'It would be great if you would let your congressman know that you hope he will support the bill.'" Michael's grin broadens. "Nearly all of them said yes."

Michael asked if they might consider dictating a message, which he would take down, and send as a telegram to Washington, DC. Almost all agreed. As Michael remembers, "most of them would actually give me money to send the telegram! Then I would head into the local town and send the telegrams straight to Wayne Aspinall."

Soon telegrams began flying into the congressman's office. "Aspinall complained bitterly, but there was little he could do. The messages were from real people with real economic interests and real addresses who all voted. He couldn't ignore them," Michael says.

Forced against his will to change his stance, Aspinall finally let the bill out of committee. While he dramatically reduced its scope, from preserving 55 million acres to only 9.1 million, the bill reached the floor and was passed and merged with the Senate bill. On September 3, 1964, President Lyndon Johnson signed the Wilderness Act into law.

Fifty-four years after its passage, 760 areas have been set aside as wilderness in the nation, encompassing 109.5 million acres in forty-four states and Puerto Rico. Oregon alone has 47 areas designated as wilderness, for a total of 2,606,439 acres. All the wilderness sanctuaries follow the same vision, as poetically stated in the law: *"A wilderness, in contrast with those areas where man and his own works dominate the landscape, is hereby recognized as an area where the earth and its community of life are untrammeled by man, where man himself is a visitor who does not remain."*

For years to come, Michael would remain involved with wilderness issues. His life, however, took an important turn a year after the passage of the historic act. In 1965 two events occurred, both of which would bring him new meaning.

Michael was offered a position by the Sierra Club to come to San Francisco and act as assistant to the president. In that same year he also married the woman who would be his partner, inspiration, and devoted wife for the next forty-one years—Maxine.

Maxine, by every definition, was an Oregonian. She was born in Oregon and loved it. Michael had become acquainted with her at state Democratic party functions, where he quickly discovered they shared similar values. Maxine had worked for Oregon senator Richard Neuberger, whom Michael respected. She was also a widow with four children.

They married shortly before he accepted the new position for the Sierra Club. Together with her four children, ages eight through fifteen, Michael and Maxine left Oregon to begin a new life in California. In less than a year, Michael became enmeshed in Sierra Club issues and was asked to be the organization's national conservation director. At the same time, Maxine was developing her own passionate interest in issues affecting wildlife, wilderness, and later, the sea.

In 1967, three years after the bill's passage, Maxine acted as volunteer executive secretary for Sierra Club's national wilderness conference. She repeated that role two years later, at the same time editing Sierra Club books for publication. She also became chairman of the Citizen NonGame Advisory Committee—the first of its kind in the country—under California's Department of Fish and Wildlife.

As Michael speaks of her accomplishments, his pride is clearly evident. "Maxine was the first to develop the idea of 'Watchable Wildlife,'" he says. The program was a novel approach to helping people realize the positive benefit that wildlife brings to a community. It sought to energize citizens to the value wildlife provides their towns and cities—both economically in terms of nature-based recreation, as well as to the importance of conserving native species in their natural habitats. "Later, Watchable Wildlife was taken up by Defenders of Wildlife and put into effect in many states, including Oregon. It is still an active program today."

His wife's accomplishments didn't end there. In fact, her greatest love of all, says Michael, were the creatures of the sea. "While we were in San Francisco, Maxine started a whale-watching group that became a powerful vehicle for whale advocacy. In due course, she became a member of the US Delegation to the International Whaling Commission, staying in that role for six years, under two different presidents."

Michael grins. "We often would joke that I was involved in saving things on land and she in the seas."

While his wife was fighting to save whales, Michael was facing environmental battles of his own. From 1965 through 1968, he took on one of the most important fights of his career: protecting the ancient California redwoods. The Sierra Club was leading the effort to create Redwood National Park, and Michael was its chief lobbyist.

"It was a long, lonely ordeal, with high consequences," he recounts. "Making matters worse, as soon as timber companies got wind that a national park might be in the future, their cutting of the giant redwoods only accelerated. I lobbied every office in the Senate and the House in Washington, DC. It was a tortuous process. The worst part was that all the lands had to be acquired. Unlike most national parks, where federal land is transferred from one agency to another, this park required buying private lands that lumber companies owned. When Congress started getting serious, timber interests began cutting even faster."

Michael was fortunate in one regard: he had met and worked with many of the officials who held key congressional positions, particularly in the Senate, during his experience with the Young Democrats in Oregon. They knew him, which allowed him entryway to their offices.

While alerting the Senate's committee staff to the dire costs of postponing action, Michael and the staff worked to develop a unique idea, a policy that would result in saving thousands of the giant, old trees.

"The staff came up with the idea of a 'congressional declaration of taking'. It was accepted. That meant, when President Lyndon Johnson signed the bill to create the Redwood National Park in 1968, those private lumber company lands were instantly condemned. As soon as he lifted his pen off the paper, they became federal lands."

The following year, Michael's work on the frontlines of the environmental movement was elevated to a new level when he became the Sierra Club's executive director. He continued his work to successfully expand Redwood National Park by forty-eight thousand acres. He also lobbied—with remarkable success—for other new national parks and wilderness areas, the preservation of Alaskan lands, and the passage of more than a hundred environmental laws.

"My poor wife; she knew I was a workaholic," Michael recollects a bit apologetically. "But we still had wonderful times together. I had a good relationship with all of the children. We were working towards the same shared values."

Maxine herself was not slowing down, nor was she in any way in Michael's lengthy shadow. Her passion to protect whales only continued to magnify. She became increasingly involved in the campaign to phase out worldwide whaling. The way she did it, as Michael shares with admiration, only proved her creativity and commitment.

"Maxine was very active in the politics behind the scenes. She was part of an effort to sign up small Caribbean countries to be on the anti-whaling side. She worked to try to get them to vote that way, to combat the votes from the pro-whaling interests on the other side. In time, she was involved with the IUCN—the International Union for the Conservation of Nature.

Maxine chaired the High Seas Project for the Oceans Commission for the IUCN.

"Maxine," he adds with tender appreciation, "was tenacious."

For seventeen years, Michael continued as executive director of the Sierra Club, during which time he was part of another significant environmental accomplishment: the passage of NEPA, the National Environmental Policy Act. Enacted on January 1, 1970, the act was one of the most important conservation achievements of the twentieth century. It codified the farsighted goal to protect the natural environment for the health and well-being of present and future generations of Americans. Its words, contained in a concise, seven-page document, remain as eloquent and grave today as they were a half century ago:

> *The Congress recognizes the profound impact of man's activity on the interrelations of all components of the natural environment, particularly the profound influences of population growth, high density urbanization, industrial expansion, resource exploitation, and new and expanding technological advances and recognizes further the critical importance of restoring and maintaining environmental quality to the overall welfare and development of man... The Federal Government, in cooperation with state and local governments, and other concerned public and private organizations (will) use all practicable means and measures, including financial and technical assistance, in a manner calculated to foster and promote the general welfare, to create and maintain conditions under which man and nature can exist in productive harmony, and fulfill the social, economic, and other requirements of present and future generations of Americans.*

Michael acted as the chief witness before the Senate Interior Committee to advocate for the act's passage. NEPA changed the course of the environmental movement in substantial ways, including requiring all executive federal agencies to prepare environmental impact statements and environmental assessments on proposed federal agency actions. Moreover, it set the stage for numerous momentous environmental policies to follow, including the Endangered Species Act of 1973.

The last twelve years of Michael's professional career took the family to Washington, DC, where he continued his environmental advocacy as the Sierra Club's chairman. In this capacity, he engaged in debates, put forth legal arguments, wrote for technical journals, and penned law review articles. After forty years of work advocating for nature, Michael retired in 2000 at age sixty-five. Maxine, he says, was ready to go home. There was no question to where they would settle. They returned to the place where their environmental advocacy had begun and where Michael, as a young boy, had developed his lifelong love for nature and desire to see it protected. Oregon.

Now, at eighty-four, that love has not dimmed. Honored with many awards for his work, Michael remains committed to conservation. He is still active with the Federation of Western Outdoor Clubs, having served as its president for two years after his retirement, and also serves as honorary vice president of the Sierra Club. Since 2005, he has focused on writing books, completing the first in 2006, the year his beloved wife and staunchest supporter, Maxine, passed away. Since then, he has authored four more.

Michael is thoughtful as he considers his life and the enduring motivation he feels to sustain Oregon.

"Why is it important to protect Oregon's native landscape? It is because it is our home. It is our habitat; other species depend on us to survive. If those of us who live here don't protect it, it will not get done. We owe that to those who will come after us, just as we have benefitted, or suffered from, what has been done in the past."

Michael pauses, contemplating experiences gleaned from working in numerous states throughout the country for nearly a half century. "Oregon is a state that stands out," he says firmly. "It is a state that leads the way. It forges new paths forward with policies that express our ideals. Of course, it is not perfect and still has places where it lags and needs to improve, but that's true of all states, and few have such an impressive record as Oregon. Why are we different? I think that good traditions of citizen activism got established early, generations ago, back to Oswald West, former governor of Oregon, and were built through the years. And I am grateful that today lots of younger people are still becoming activists."

Michael acknowledges he has seen many heartening wins. He has seen some painful losses. He is well acquainted with the difficult road he knows lies head. But one thing is certain. He will never relinquish his hope.

"Everything that has gotten done in the past is because somebody cared. They cared passionately, often devoting much of their life to it. New issues are emerging all the time. Attacks are being made to roll back achievements of the past. It will be never ending. It takes new generations, succeeding the others, to pick up the mantle and to carry the cause forward. It requires people who believe in themselves and to understand that they can make a difference."

Michael smiles, with the wisdom of a life well-lived reflecting in his eyes, as he shares a secret he knows.

"Never underestimate the influence that an individual can have."

Those who love Oregon often find a special spot that is theirs to believe in—one that grounds them, one they are willing to put their heart and soul and time to care for.

————————

Betsy Johnson

18

Her Mother's Daughter: Steward of the Metolius
BETSY JOHNSON

A fisherman prepares to cast a line in the early morning light of Central Oregon, delighting in the pine-scented ground and cold, pristine springs forming one of Oregon's clearest, most prized rivers. He is not alone in cherishing the Metolius River. Thousands of Oregonians count it as one of the country's greatest streams, including one woman willing to risk her entire political career to do everything she could to protect it in perpetuity.

❧

Picking up litter along the river might not have been exactly what Betsy wanted to do as a child, but she had little choice. She came from a family where public service was valued and practiced. Even at a young age, Betsy realized that the place where she was lucky enough to grow up—between Redmond and Camp Sherman in Central Oregon, where her family owned 160 acres along the Metolius River—was special to her parents. Its stately pines, green meadows, and especially its glorious headwaters were worthy of protection, she understood from observing the actions of her father and mother.

"Even when I was little and my mother planned missions for 'cigarette butt patrol,' I wondered how could people go to a place that magnificent and throw garbage on the ground. I couldn't fathom it," says Betsy. "I grew up indoctrinated with the belief that we were stewards of the Metolius and that it was one of the unique places on the planet."

Her childhood, surrounded by such beauty, instilled in Betsy "a love affair with Oregon that is in my genes," she attests. The upper Metolius is a state and federally designated wild and scenic river, set aside for its magnificent qualities. The federal Wild and Scenic Rivers Act, signed into law in 1968, selects rivers across the nation "to be preserved in free flowing condition, and they and their immediate environments shall be protected for the benefit and enjoyment of present and future generations." The splendor of the Metolius is legendary across the nation, with its sparkling clear, very cold waters, teeming with rainbow, bull, and brown trout. Arising from plentiful springs at the base of Black Butte, a remnant cinder cone, the river winds through grassy meadows rich with plant life and watched over by glorious stands of old-growth ponderosa pines. River otters, beaver, mule deer, elk, black bear, and cougar roam and share the wild landscape.

Betsy absorbed the feelings her parents had for the Metolius as her own. She acknowledges that her mother and father played a large role in her development. They showed her what it meant to advocate for something you believed in. They also gave her the courage to follow her interests.

"My sister and I got our airplane licenses early," Betsy relates. "My father, Sam, had encouraged it. After I received my pilot's license, I went on to get my helicopter license. My sister became a nurse and I went to law school and graduated. I was in the midst of getting ready for the bar exam, when I got a call from Bell Helicopters saying, "You want to come represent the United States on a flying team that's going to the [then] Soviet Union?"

It was an easy choice for adventurous Betsy. She was off to Europe. For nearly twenty years, from 1977 to 1994, Betsy flew helicopters. In 1978, she founded a helicopter company. A year later, she met John—another helicopter aficionado—who performed maintenance on helicopters and whom she would later marry in 1986.

Betsy flew to fight fires, to survey for peregrine falcons, and to take Disney moviemakers in the air to film for blockbuster productions. Her sister, bitten by the same flying bug, gave up the practice of nursing to join the US Unlimited Aerobatic team, where she flew all over the world in aerobatic competitions and became a corporate pilot.

Through all that flying experience, Betsy learned to see Oregon from different perspectives. She soared over the verdant Douglas-fir forests of the Cascade Range; she hovered over the High Desert of eastern Oregon and up and down the Columbia River Gorge along the Oregon and Washington borders. She saw the Owyhee Desert and the majestic Wallowa Mountains. What she saw stayed with her.

"We need to take care of this place called Oregon. We need to take care of its physical attributes. We need to find the things that unite us and not the things that divide us."

The beauty she viewed from the air deepened her affection for all of Oregon and especially the Metolius River. It also strengthened a belief—one that she saw evidenced in both her parents' actions for over half a century—that the Metolius was an extraordinary area worth fighting for.

Betsy's attachment to the place came naturally enough. Her father, Sam Johnson, had come to Central Oregon in the late 1920s to work as a lumberman on forest properties owned by his family. Residing in Sisters, Sam quickly became captivated by the Metolius River. In 1935, he purchased 160 acres, including the headwaters, from his father. He bought a house from a Sears, Roebuck and Company catalog and placed it strategically in a meadow he had learned was to be the site of proposed power lines.

"Those power lines then had to go someplace else, and that magnificent, iconic vista was preserved," explains Betsy proudly.

While Sam envisioned only staying in the area for a little while, he ended up living in Central Oregon for the rest of his life with his wife, Becky. When Betsy was nine years old, the family built a summer home upstream from the old house. And while the 160 acres were still privately held by the family, they allowed public access to the beautiful Metolius headwaters. Motorists drove on a path that crossed their property to look at the view.

Before long, more and more visitors came. They drove their cars and came by Greyhound buses. Not only were the Metolius Springs and the vegetation surrounding them becoming trampled, the site was becoming transformed by blowing dust.

"This was a real problem. People used to be able to drive literally right down to the water," Betsy explains. "My parents were constantly aware, and concerned, about the fragility of the springs. There are springs all along the upper part of the river. One main spring originates on our property, but there are other, little quiet spots along the shore, where water bubbles up as it comes into the river. That's part of the mystery of the Metolius. Nobody really knows where it starts."

And there were other issues. The soil of the Metolius basin is very fine, the texture almost of cosmetic face powder. When cars began driving close to the water to get a better view, they spewed damaging dust into the river. Betsy's parents tried several different techniques to keep automobiles away from the headwaters. They rolled large sections of logs to block travel. They moved logs farther back to create a parking area. The vehicles just went around the barriers. Before long, the denigration of the place she loved became too much for Betsy's mother.

"The final straw for my mother happened one day when she was on her cigarette patrol. There was a woman, an artist, setting up her easel and paints by the river. My mother spied her, and watched her beginning to paint a scene of the spectacular view. At that moment, a great tour bus came into what was ostensibly the parking lot created by the logs that my parents had rolled in to keep people back. As it stopped, clouds of fine powder engulfed the woman and her paints in fine dust. My mother came back to the house, furious.

"She told my father, 'This has got to change. We've got to do something about this.' And that's when the idea to preserve the Metolius was hatched."

Betsy's parents contacted the US Forest Service about moving the road up and away from the Metolius springs. In trade for that concession, they would entrust the land around the headwaters, the viewpoint, the trail, and the parking lot to public ownership. There was one condition, though—Betsy's parents insisted that the path to the springs be paved. Long before the American Disabilities Act (ADA) of 1991 was an issue, they maintained that less able-bodied folks and elderly people needed a way to get down to the headwaters.

On July 10, 1971, the Johnson family dedicated the priceless Metolius River headwaters to the US Forest Service. It would be the first scenic easement ever to be processed by the Department of Agriculture.

Fifty years later, Betsy, who still owns the land around the deeded property, continues that partnership with the US Forest Service. Over 250,000 people visit the viewpoint annually. The fight to protect the headwaters and basin, though, didn't end with the donation. While its scenic easement guaranteed that nothing would ever be built there, the responsibility to be caretakers of the larger Metolius basin was aggressively upheld by the family. With each new threat, Betsy saw her mother rise to the occasion to preserve the character of the Metolius for future generations.

"One of the first campgrounds after our place is on federal property," Betsy says, recounting an early battle. "At one point, the US Forest Service came up with a proposal to develop it in a fashion where huge motor homes could come down. They planned to build showers and laundromats, all the rest of the stuff. My mother said, simply, 'Over my dead body you're going to do that!' So my mother went to war."

Becky Johnson hired a Washington, DC, lobbyist to help change the US Forest Service's plans. With deep affection in her voice, Betsy explains why she took on the fight. "My mother believed, as I do, that people come here for the experience of the grandeur of nature, for the quiet, and the solemnness of the nights. They don't come to have the engines of giant motor homes running all night long. My mother won. It is all walk-in tent camping now. And it will stay that way as long as I'm alive."

The walk-in campground was just one of her mother's legacies. In the late 1980s and early 1990s, populations of the native Western pine beetle erupted in Central Oregon. They were devastating the beautiful ponderosa pines that her mother called the "cathedral trees." When beetles hatched, they would fly to the bark of the trees, bore down to the cambium layer beneath the outer tree bark, and kill them.

This did not sit well with the Johnson family. "My mother set out to save those trees like she was saving drowning people. We all worked to wrap the old giants with fine mesh wire, up to about twelve feet high, so the beetles couldn't get into the tree. We didn't do the entire 160 acres, but we did cover all the big old yellow bellies—as we called those giant ponderosas—close in. After they were wrapped, my mother continually checked up on them. And the good news is, all but a handful managed to survive."

Betsy becomes thoughtful at the remembrance. "Growing up in my family, there was always a stewardship ethic that suffused my relationship with that place. My own confrontation to save the Metolius would come later."

Arguably, it would be the most ferocious fight of all.

In the 1990s, twenty years after the Johnson's had deeded the headwaters to the US Forest Service, Betsy's active life in aviation life took a new turn. She sold her interest in the helicopter company and became a lobbyist for the aviation community in Salem, Oregon. Soon she was recruited to become the aeronautics administrator for the state of Oregon, "which is a fancy title," Betsy says dryly, "for a mid-level bureaucrat." A seat opened in the legislature, and she ran for the Oregon house in 2000 and won.

Betsy was reelected twice more. Ten days into her third term, though, she was offered a vacated state senate seat. By unanimous vote of five sets of county commissioners, Betsy was appointed to fill out the term. Since then, she has continued serve as state senator, representing a large portion of the Oregon Coast, from half the city of Tillamook north to Astoria and up the Columbia River to Sauvies Island.

Early in her tenure as senator, a new threat was posed for the Metolius—one no one had expected. It became a raucous fight in the legislature, and Betsy found herself squarely in the center of it. Two massive developments, representing thousands of houses, were slated to be built in the Metolius drainage. Both would have heavy draws on the Metolius River, with the potential of changing the watershed forever.

"You can only stick so many straws into the aquifer—the developments, all those golf courses getting water, toilets being flushed, showers being drawn—and then one day the magical, mystical Metolius is gone, and everybody goes, "Oh God, I wish we hadn't done that."

Betsy could not let that happen. Her father, Sam, had been gone for twenty years. Her mother was ninety-three and in poor health. Both had been warriors for the Metolius. It was her turn now.

"My argument was, if I lose my political career defending the Metolius, so be it."

State senator Ben Westlund, another advocate for the river, proposed a bill to try to create an "area of critical concern" in the Metolius that would prevent the developments from going forward. Under the original legislation, Senate Bill 100, special places could be set aside from development because they were of such high value. This was one of those cases. Betsy fought for the passage of the bill ferociously. Before long, most of developers and other pro-growth advocates pointed their hostility toward Betsy. It generated nasty press.

"They said 'she's just doing this to protect her interests.' Well, bullshit!" Betsy exclaims. "I was doing it to protect the basin!" And she was doing it in honor of her mother, who passed away during the middle of the legislative melee.

The bill moved through the system. The vitriol Betsy endured grew more heated. Betsy admits there were rough days and nights for her personally, but nothing would change her absolute resolution that she would see it to the end. She was her mother's daughter, after all.

There were several nip and tuck votes on both the floor of the house and the senate. Heavy lobbying ensued on both sides of the bill. Campaigning was bellicose, confrontational, and unrelenting. In the end, though, the bill passed. On July 16, 2009, the Metolius basin was designated an Area of Critical Statewide Concern (ACSC), a first-of-its-kind classification in the Oregon landscape. The passage of the bill restricted development in the Metolius headwaters in southern Jefferson County. Large-scale development of the pristine basin would never be allowed.

"I believe the bill passed," reflects Betsy philosophically, "because of the growing recognition that there are precious places in Oregon that are worthy of saving. Make no mistake," she adds, "I believe that there are opportunities and places that could and should be developed and should be the locus for jobs and opportunity. I grew up in a timber family involved with forestry, I'm the fourth generation, and I understand that there can be wise use of resources. I saw my father's involvement in the timber industry in Central Oregon, and observed an ethos about sustainable harvest and protection. But there are other places that need to be stewarded and protected. And they need to be guaranteed in perpetuity that they're going to be the way they are."

Not all people are in a position where they can leave a place as a legacy gift, as Betsy's parents were able to do. The Johnson family had a unique opportunity to deed the Metolius headwaters and its spectacular views for everyone to enjoy. But every single Oregonian, Betsy believes, can be helpful in thinking about how to preserve those really special places that are worthy of that protection.

"It isn't just the tangible preservation of real estate, it also is the advocacy behind it, and the stewardship. Those who love Oregon often find a special spot that is theirs to believe in, one that grounds them, one they are willing to put their heart and soul and time to care for."

That place, for Betsy, will always remain the Metolius.

"I had many choices of where I could live, but the pull of Oregon was always gravitational," she admits with emotion. "This is where I grew up, this is where I live. This is where my family history is, this is where my parents are buried. This is where I've chosen to make a commitment to public service. I know this place, this place knows me.

"This is home."

Life is the only wealth,
and it springs from the land.

————————————————

Tom McAllister

19

Take the Children Back Outdoors
TOM MCALLISTER

Gifted storytellers play an important role in culture, translating history, events, and places for us, making them come to life. For nearly forty years Oregonians had such a storyteller, whose words illuminated the state's natural history. His newspaper narratives sparked a new appreciation of Oregon's rivers and wild places, its fish and wildlife, and called many to action to protect them.

∽

A deep love of birds and a lifetime of conservation efforts began for Tom McAllister when he was twelve years old. It sprang from an experience offered in Oregon over eighty years ago that inspired him to care for the natural environment surrounding him.

It all started at school.

In 1938, the year Tom's interest caught fire, Portland schools offered a program entirely unique to Oregon and the country: Nature Study. It was a yearlong course incorporated into the grade school curriculum. Teachers were instructed by a noted scholar, Professor B. A. Thaxter, an active member of the Oregon Audubon Society. Dr. Thaxter held educational sessions at Lincoln High School in Portland and followed them up with Saturday nature hikes.

At this time, Oregon led the nation in its involvement in nature education, with much of the credit due to the influence of the Oregon Audubon Society. Established in 1902, the Portland chapter was among the first in the United States. The initial support of the Audubon Society came from President Theodore Roosevelt, himself an ornithologist. As Tom recounts,

"Theodore Roosevelt reveled in the robust outdoor life, and was known to puzzle White House staff by walking outside to stand motionless for long periods under the trees."

In addition to Portland's active chapter, other Oregon Audubon groups—those headquartered in La Grande, Salem, Baker City, and Joseph—supported environmental programs. Local chapters partnered with school districts to bring natural history science programs to classrooms. Students became involved throughout the year with nature observation and hands-on field collecting, and they reported their discoveries to their class.

This progressive type of educational opportunity expanded the horizons of thousands of children, opening their eyes to the natural world around them and to its significance. For some, like Tom, it set their life's course.

"It was a wonderful time for science education," Tom explains, "a convergence of kindred souls. The Oregon Audubon Society had a strong purpose; at the same time, the nation was benefiting from having had a president who set in motion the conservation movement."

The vision of the Oregon Audubon Society, Tom relates, was clearly stated: "To protect the wild birds and animals of the State of Oregon, and by literature, lectures, and all other available methods, to disseminate knowledge and appreciation of the economic and esthetic value of wild bird and animals." In similar fashion, Teddy Roosevelt had modeled the same commitment. "President Roosevelt set aside 230 million acres--a public lands legacy that included national forests, parks, monuments, and refuges."

As a young student, Tom took both messages entirely to heart. They blazed a purpose for his life has persisted for over eight decades. After that class, Tom was hooked. It set his life's trajectory and he never looked back.

Starting in high school, Tom took jobs at the Fremont National Forest and the Malheur National Wildlife Refuge. As Tom describes, Malheur Lake was one of three federal refuges in Oregon that President Roosevelt had decreed by executive order in 1908. The two others were Lower Klamath Lake and Three Arch Rocks—the latter the first bird refuge in the West. All three sites abounded with waterfowl.

Tom thirstily absorbed rich information during these formative years from a host of early naturalists with whom, he readily notes, he was privileged to work. His mentors remain highly regarded for their groundbreaking efforts in conservation: Stanley Jewett, US Biological Survey; John Scharff, manager of Malheur National Wildlife Refuge; and Leo Simon, Native Plant Society, among others. The collective wisdom they passed on to Tom furthered his drive to work with wildlife and to study the exciting natural environments of Oregon.

Then all nature study came to a halt. World War II broke out and, as for many young men, Tom's schooling was interrupted. He was called to duty as a US Navy Corpsman. Because of the great need, he enrolled to become a medic. After completing training, he embarked on a hospital apprenticeship at Camp Adair, near Oregon State College in Corvallis, before shipping out for the Philippines and the Pacific Theater.

Liberty breaks during training, says Tom, were scarce and therefore precious. For his first one, he decided to hop on a bus and head into Corvallis. He noticed a few co-eds already seated on the transport. One, a redhead, grabbed his attention. There was something about her he vaguely recalled. Tom knew he didn't have much time. Moving to sit next to her, quickly he seized his chance.

"Didn't you go to Grant High School in Portland?" he asked the pretty girl.

She replied that she had. Talking a bit further, they discovered that she had been two years ahead of Tom in school. Her name, she said, was Barbara.

Before they disembarked, Barbara told him there was to be a dance that night at the Oregon State Memorial Union. She said Tom would likely know some of the people there. Kindly she offered to meet him there and locate the girls from his class at Grant that he might enjoy visiting with.

Tom thanked her and, later that evening, headed over to the dance. He saw her with several old classmates that Barbara had generously rounded up. None of that was important to Tom. He was already smitten—by the redhead.

"I danced the whole time with her," he remembers with a smile. "I walked her back to the Pi Phi Sorority house that night. And every liberty leave I had for the months I was stationed at Camp Adair, I went to see her. After I shipped out, I had an unbroken correspondence with her until my return in 1946."

They married on December 18, 1948, after the war was over, while both were enrolled at Oregon State.

Tom relished being in college. Summers between coursework at Oregon State, he adds, were equally stimulating. During those months, Tom took a job managing a twelve-mule pack string while surveying and stocking multiple Cascade Mountain lakes with fingerling trout. A large part of his love for Oregon wilderness was formed during this time.

"In the 1940s, I could look out over absolute unbroken, forest country mixed with mountain meadows. There were no roads whatsoever; it was all defacto wilderness—a wilderness that covered the entire Cascade Mountains in Oregon. My surveying job required I go up on top of every lookout, in particular Maiden Peak, Diamond Peak, South Sister. In those days we didn't have Global Positioning Systems; it was primitive mapping—locating ourselves, orienteering, tying things in with the different geographic features to get a feel of the whole country." Tom regrets what happened later. "It all went so fast. The speed at which we went through and logged the entire Cascade National Forest was staggering, except for the highest country. Great battles were waged by early conservation groups to try to retain what little wilderness remained."

Tom himself was involved with several such efforts. Two were remarkably successful, to his lasting satisfaction: the fight to save Waldo Lake, a pristine body of water in Central Oregon's high Cascades, and the effort to create and then expand the Eagle Cap Wilderness in northeastern Oregon.

"These were huge accomplishments, just getting low elevation land preserved, the ponderosa pine and tamarack stands. We also were able to protect the entire Minam Watershed in the Eagle Cap Wilderness. This achievement saved one great river, which is nearly completely intact from its very headwaters down to its juncture with the Wallowa River and on into the Grande Ronde River. It remains today a watershed that is entirely roadless, except for the very lower end.

"We need places like this," Tom affirms. "There is something about openness and space that in itself gives a feeling of freedom—whether its plains or prairies or mountains."

After graduating from Oregon State College with a degree in wildlife, Tom was hired by the Oregon Fish and Wildlife Commission and worked for them for three years. He was then offered a job that would recalibrate his life's trajectory. The *Oregon Journal*, a highly respected newspaper published in Portland, proposed that he write a column for them. The opportunity would allow him not only to study the fish and wildlife of the Northwest, but to write about them.

In 1953, Tom became the first full-time outdoor writer in the region. It was a position he loved, and held for the next thirty-nine years. The travel, research, and writing shaped him as an expert like none other on the outdoor life of Oregon. In turn, he helped all those who read his column, Wide Open Spaces, understand the natural wonders of Oregon. The column ran three times a week with a full outdoors page each weekend. His articles followed the seasons, with stories on camping, fishing, and hunting. In winter, he covered skiing activities around the West with a Ski Scope column.

Tom's stories were the stories of Oregon. He reported on the natural history of a place, its geology, anthropology, birds and other wildlife, and the nature of the people who lived there.

More importantly, his power of communicating what was special about Oregon helped people fall in love with significant spots within it, and from that, to wish to conserve them.

Tom's dedication to conservation issues grew at the same time his fund of knowledge of Oregon's wildlife and natural areas continued to expand and burgeon. With his friend and fellow biologist, David Marshall, he wrote scientific as well as journalistic pieces that aided in the 1964 creation of the William L. Finley National Wildlife Refuge, located in the Willamette Valley south of Corvallis. He played an integral role in the fight to save Tillamook County's Kilchis Point—a rich coastal ecosystem abounding with waterfowl and a place he had loved since childhood—from development. As Tom recounts, Kilchis Point was not only valuable as wildlife habitat, it was also imbued with rich historic significance. It was the spot where Robert Gray, captain of the first US ship to circumnavigate the globe and explorer of the Columbia River, landed and placed the first American flag on the West Coast.

Tom was the founding member or president of numerous conservation organizations, including the Oregon Chapter of The Nature Conservancy, which he helped establish and then served as president for many years. He was president of the Flyfisher's Foundation and active on the boards of the Oregon Wildlife Foundation and the Izaak Walton League, becoming chairman of the latter. He was also one of the original founders of the Oregon Parks Foundation and participated in that philanthropic conservation organization for more than thirty years. Always mindful of the tremendous impact it had on his life and future, Tom remained committed to the Portland Audubon Society.

His expertise on Oregon's land and rivers was noticed in political corners as well. When the Wild and Scenic Rivers Act became law in 1968, then governor Tom McCall asked Tom to be president of the committee that nominated Oregon rivers to be protected under the legislation. Tom keenly remembers the delight he felt the day that Governor McCall signed a law naming Oregon's Rogue River as one of the original eight rivers designated as national wild and scenic waterways in 1968.

Tom is straightforward about the core of his motivation to write about and devote his life to conservation issues. It is based on a philosophical conviction held jointly with Barbara.

"The protection of Oregon's natural resources is critical to the state's future. Dramatic losses are ahead if rampant and unplanned development is not addressed. Life is the only wealth, and it springs from the land," Tom attests. "Preserving areas takes people back to their roots. America is relatively young; our roots are not very deep; but we do have this attachment to the land, and it grows. It is important to have places where one can go to experience the real foundations of Oregon."

He adds, though, a note of caution. "It is becoming harder and harder to find these special spots because of modern technology. It is therefore critical to keep these special spots true to what they have always been, rather than to modernize them. Otherwise they erode away and become a Disneyland type of experience."

Tom finds encouragement from the proliferation of small, nonprofit conservation groups in Oregon, bands of citizens who have come together with goals to support preservation at the local level. The North Coast Land Conservancy, Southern Oregon Land Conservancy, Wallowa Land Trust, and Wallowa Resources, he says, are just a few examples of people working to protect their native landscape, whether a specific watershed or geographic unit.

"To me, these grassroots groups evince this desire to experience what the land originally was, and what through preservation and good steward-ship can be again. We have had wonderful citizen representation in Or-egon, as far as this longing to save or recapture some of what we had at the beginning. Senator Mark Hatfield was out in the forefront to making new wilderness areas in Oregon. Senator Bob Packwood created the Hells Canyon National Recreation Area. Governor Vic Atiyeh worked to save the Deschutes River. Governor Bob Straub preserved our coastline by stop-ping the construction of the highway from going straight down the coast. Straub was a Democrat; all the others were Republican," says Tom, "but that's when we had a different breed of Republican."

In 1993, at age sixty-six, Tom retired from his long and celebrated writ-ing career at the *Oregon Journal* and later for the *Oregonian*. It did not stop him, though, from continuing his mission to educate others about Oregon and the natural world. For the next fifteen years, he was a popular natural-ist/historian for Lindblad Expeditions cruises in Alaska and on the Colum-bia River. Today, at age ninety, his interest in nature and passion for birds has not waned. He remains rooted and devoted to Oregon.

"There is no other state in the country quite like Oregon. It is that whole feeling of distant great horizons. What Barbara says is true for me: 'I have always felt blessed that our families were Oregonians.' Because our areas are exceptional, we have sensed the need to protect them. We need to always treasure Oregon's wilderness that sets us apart."

But how do we do that? How do we keep that feeling alive? Tom re-sponse comes swiftly. He returns to his original inspiration and a belief about what is essential for the future of Oregon if it is to remain the place he loves.

"We have got to have those outdoor programs back," he says, unwaver-ingly. "We must take the children back outdoors."

Just do it.

Be persistent and you can succeed.

Just do it, and never give up.

Have the courage to get it done.

—————————

Nancy Russell

20

She Said Yes

NANCY RUSSELL

as told through her son's and sister's eyes

Sheer basaltic cliffs rise up hundreds of feet. Dozens of waterfalls grace the hill-sides. Golden hilltop prairies are strewn with blazing yellow arrowleaf balsam-root, while other colorful wildflowers, some found nowhere else in the world, lie tucked within old growth forests. Through it all, one of the nation's outstanding rivers slices a prominent course dividing two states.

To protect this vast landscape—the Columbia River Gorge National Scenic Area—would require an act of Congress in a time when national conservation legislation had ground to a halt. It would also take the bravery of one woman who would lead the grassroots effort to save it all the way to the vice president of the United States.

∾

Oregon native Nancy Russell was a woman who could climb mountains and name every wildflower in the Cascades. She could charm a vice president of the United States and inspire others to see the incredible beauty around them. She could bring a spirit of fun into everything she undertook.

She could also endure blistering personal attacks for a cause she deeply believed in, and, in time, lead an entire nation to protect one of the world's most significant natural landscapes: the Columbia River Gorge.

"Nancy was the strongest woman, and the most courageous, I have ever known," says her sister, Betsy Smith. "The essence of her happiness came through the beauty of nature. She had a persistent quest for beauty and intense interest in being outdoors, which all started, I think, on our family's farm."

Born in 1932 at the height of the Depression, Nancy spent her earliest years on her grandparent's farm. Her father, when finding himself suddenly unemployed, had moved to his wife's family's home in Dundee, Oregon. The farm was perched on the top of the hill, surrounded by beautiful fields and natural woodlands. To Nancy, as a very young girl, it was a wild paradise.

By the time she entered grade school, Nancy's father had found a new job, and she moved with her parents and older brother to Portland. The farm, however, remained in the family and would be a touchstone for her as she adjusted to life in the city. At first, the family moved numerous times into different rentals until Nancy's father found more stable work and they settled down in the southwest hills, where Betsy was born, seven years after Nancy, in 1939.

The disruption caused by moving, however, held a positive influence for Nancy, Betsy recounts. While the family did not have much in worldly goods, they were secure and happy. Doing more with less, Nancy learned self-reliance, how to entertain herself, and the art of being self-taught.

The simple lifestyle of their childhood also included something more: treks to the wilderness. To Nancy's great and lasting joy, her family had a cabin on Elk Lake in the Oregon Cascades, which had been in her family for generations, and where would they spend every summer hiking, and in winter, on skis.

"Elk Lake was hugely formative for my sister," affirms Betsy. "She was keenly aware of all the wilderness, wildflowers, and wildlife around her, and with her great sense of fun, wanted to share it with me. Even though I was seven years younger, we were very close. No two ways about it! We were soul mates.

"Sometimes, though, my sister would ask me to do things I didn't feel brave enough to do," she adds honestly. "When she knew I was scared, she would merely encourage me. 'Just do it!' she'd exclaim. 'Don't give up.' Looking up to my older sister, whom I adored, I would always try, even if I didn't always *like* it."

Betsy distinctly remembers an event when she was nine years old and Nancy was sixteen. Early one morning at the family cabin, Nancy came into her room to wake Betsy up. It was raining outside, and Betsy could hear thunder.

"Come on, let's go!" Nancy said. "It's a perfect day to climb up the glacier!"

"But lightning can kill you can't it, if it strikes and you are not under a tree?" Betsy recalls saying, feeling afraid. Her sister, however, was fearless.

"Don't worry about that, Betsy. Stop fussing; come! It will be a grand adventure!"

Betsy relates her sister had a poor sense of direction, but that did not give her pause. With positive energy and overflowing good nature, she took her sister high into the hills, up to the snow fields along the woods.

"We will follow the blazes," Nancy said, joyous to be outside amidst all the wild beauty she loved. "It's not dangerous. We will get up to the top of the mountain."

Betsy followed her sister, and they made it to the top of the glacier, exhilarated. Once there, Nancy, seeing that the light of the afternoon was dimming, said they needed to turn around. "We don't want to be caught in the dark, so get out your flashlight, Betsy. You *did* remember to put it in your pack, didn't you Betsy?"

Betsy shrugs her shoulders, not remembering if she had or not, but it really didn't matter too much. Nancy, says Betsy, had a gift of laughter. "Just being around her, you felt you could succeed."

They made it back just fine.

Nancy graduated from Catlin Gabel High School, a small private school in Portland, where she had attended on scholarship, then attended Scripps College in San Diego for four years, also on a scholastic scholarship. She majored in English literature. After graduation, she worked for several years as a librarian at Reed College in Portland. In 1957, she married Bruce Russell, another native Oregonian who shared a love of history and nature.

Betsy explains it was Bruce who introduced Nancy to conservation and environmental activism. He had been involved in environmental issues since his college years at Stanford University. An early member of the Sierra Club, at the time a fledgling conservation group based in San Francisco, Bruce had advocated for the federal purchase of the remaining redwoods, which later became Redwood National Park. After college, he moved back to Oregon. When he and Nancy married, they settled down in the home where he had been raised in southwest Portland.

Soon children arrived. Home became Nancy's joy and focus. As her son, Aubrey, details, "She made curtains for the house, did her own ironing and cleaning, collected S&H green stamps for redemption at the grocery story, canned peaches and plums for winter. She planted her own seeds, maintained a large vegetable and flower garden, and mowed her own lawn. She was tender in her love of her children and through it all could not hold back from having fun. Her sense of fun extended to everything she did."

Aubrey, the youngest of the five children, had three older sisters. His older brother, Hardy, had died at age two from spinal meningitis, five years before Aubrey was born. It was, as he and Betsy both relate, the great tragedy of Nancy's life.

"But she courageously faced that too," says Aubrey, speaking softly. "And, in a way, it is central to my mother's story. Because of the depression she was under, she had to take her doctor's advice to get out of the house. She worked at the World Affairs Council and volunteered at the Oregon Historical Society. That forced her out into the community, and she found that there were other things that could give her life purpose and meaning beyond everything purely domestic. She discovered she could sink her energy into her kids and other great things."

Nancy's work and involvement in the two organizations expanded her world and imagination. Over time, her stamina and energy revived. She had two more children after Hardy, including Aubrey. As the children grew, she took them on the sorts of expeditions she loved best: hiking, canoeing, swimming, and camping adventures.

Those migrations began to center on the Columbia River Gorge east of Portland. The stunning landscape of the gorge captivated and inspired her. She loved the steep basaltic cliffs that towered over the Columbia River's shoreline. She took pleasure in the more than thirty hurtling waterfalls that graced the precipitous slopes of green and rock. Acres of old-growth forests of Douglas fir and cedar filled her with reverence. Spectacular views of snow-clad Mount Hood, Mount St. Helens, and Mount Adams, seen from many vistas, renewed a sense of wonder and awe. She gathered up her family and traveled to islands for swimming, to the numerous falls for picnics, to the gorge's eastern plateaus for campfires and to roast hotdogs.

It was only natural that Nancy soon found herself delighting in the wildflowers of the gorge. Over time, that interest would grow to become one of her greatest passions, and captivate her for the rest of her life.

Betsy smiles as she recollects how the family "wildflower hikes" grew in size and scope.

"Nancy always loved wildflowers; that interest started at Elk Lake. It was, then, a paradise. We were surrounded by flowers, particularly up in the Three Sisters Wilderness. When Nancy's children were young, they'd go on short wildflower walks. Then, the hikes became longer. And longer. She joined the Native Plant Society and the Portland Garden Club. Soon she became an expert. It was all self-taught."

Betsy laughs. "My sister always had one of those magnifying glasses dangling around her neck. She would forever be squatting down on her knees to look and just be sure of the genus and species."

Wanting to share that fascination, their interest in history, and their values with their children, Bruce and Nancy began researching original pioneer journals and studied the route of the Oregon Trail from St. Louis, Missouri, to Oregon City, Oregon. For several summers, with their kids in tow, they traced the Oregon Trail from Missouri to Oregon, taking pictures all along the way. From these, Nancy created informative slide shows and educational programs for the Oregon Historical Society that she presented in Oregon schools, clubs, associations, and even the state legislature.

It was at one of these presentations that something happened that would change the course of the rest of her life and the history of Oregon.

A well-known landscape architect Wallace Huntington attended a wildflower program at the Portland Garden Club, where Nancy was the featured speaker. He had heard about her educational series and wanted to see one for himself.

"He told me later, 'Your sister is a remarkable person!'" Betsy remembers. Thoroughly captivated, Huntington shared his discovery with his good friend, John Yeon, eminent Oregon architect and preservationist. Huntington knew that Yeon had been working for decades to protect the Columbia

River Gorge from development. The place was deeply special, and Yeon ardently felt it should become a national park. He had been faced with frustration after frustration, however, his efforts repeatedly countered with "it's an impossibility."

After years of futile attempts, Yeon had grown tired and pessimistic. Looming, too, was a proposal for an Interstate bridge to be built connecting Portland and Vancouver (Washington), which, as John Yeon knew, would bring suburban and industrial sprawl flooding into the gorge. Subdivisions were already being platted on the bluffs. His love of a place that he called "the most noble landscape in the United States" remained fierce, but his hope of ever protecting it was rapidly fading.

Huntington changed that when he confided to his friend that there was a person that might be just the one to take up Yeon's longstanding yet dimming charge. Her name was Nancy Russell. Intrigued, Yeon wanted to see her for himself.

Introduced by mutual acquaintances, Yeon invited Nancy and Bruce for dinner at his shoreline residence along the Columbia River in the Gorge. The beautiful waterfront landscape, located on the Washington side, lay in the heart of the gorge, directly across from magnificent Multnomah Falls. Yeon had purchased the seventy-five-acre property that he affectionately named "The Shire" in 1965 when it was threatened from possible industrial development. In the fourteen years since becoming its owner, the architect had masterfully designed walking paths and vista points, wending through green meadows and wetlands, while keeping the integrity of the native landscape intact. In time, the entire property would be permanently preserved, becoming part of the University of Oregon's John Yeon Center for Architecture and the Landscape, to inspire present and future generations of Yeon's visionary design and lifelong work to preserve the landscape of the Northwest.

The Shire was Yeon's most treasured place, and he wanted to show it to Nancy at its best. The circumstances had to be perfect, to introduce her to the idea that was brewing inside him. Yeon cancelled the dinner event two times because of cloudy skies. At last the conditions were right, and on a late summer evening in August 1979, Nancy and Bruce arrived at the Shire.

"There was a rising moon across the river, right above Multnomah Falls," says Aubrey. "There was the pink evening light on the cliffs. My mother said later, she was completely awed. The evening was magical."

It was at this moment that John Yeon, recognizing that Nancy's ideals of beauty were resonant with those deepest in himself, asked her if she would lead a national effort to protect the Columbia River Gorge and save this majesty.

Nancy, overwhelmed with surprise, did not answer right away. The commission, realistically, was unimaginable! Formidable! She had children still at home. She had no knowledge of politics or legislation. Not wishing to offend her host, she told Yeon she would think about it.

For several weeks, Nancy grappled with the idea. Then one day, she travelled to Crown Point, a scenic, basaltic promontory on the Historic Columbia River Highway. Standing at the windswept overlook, Nancy gazed up and down the river, deep in thought, taking in the outstanding, extensive gorge views everywhere around her.

"My mother saw those lands across the river, on Mt. Pleasant, in Washington. There were fertile agricultural lands and fields of unmarred prairie, completely unprotected by any sort of zoning or protection from development. Where she stood was squarely in the center of the iconic view of the Columbia River Gorge," says Aubrey. "She knew then, this beauty should not be ruined. She knew, if she did not act, and these lands became developed and spoiled, she would never forgive herself."

Betsy remembers her sister telling her something more. "In her forthright way, she told me, 'Betsy, when the time comes for me to cash in my chips, I won't be able to, until I've saved that!'"

Nancy agreed to take the mantle from John Yeon. And then the real work began.

Yeon had always envisioned the strongest safeguard to preserving the Columbia River Gorge would be to set it aside as a national park. Unfortunately, as Nancy soon learned from Oregon Senator Mark Hatfield, that was not a possibility.

"Senator Hatfield knew and supported John Yeon and everything that he stood for," explains Betsy. "But he was also incredibly knowledgeable about legislation and politics. He told Nancy there was not a prayer of getting it nominated as a national park; there were just too many candidates."

Nancy was undeterred. Instead of taking that reply as a "no," she entreated that the area between Troutdale and The Dalles be put under conservation in some other fashion, to ensure its protection in the public domain. Senator Hatfield agreed.

"The Senator told my sister that he thought it might be possible to create a national scenic area under the auspices of the US Forest Service," says Betsy. "In fact, he was strongly in favor of the idea."

Even more, he volunteered to help. He saw Nancy's determination and commitment. She held high standards, was direct and efficient, without forsaking the idealism that motivated her. "He told Nancy, "Don't worry about the legislation; I'll take care of that. You just do the grassroots. You bring me the grassroots votes and support and I'll do the legislation.""

With that encouragement began a trial that would test every drop of Nancy's resolve. It would, as Betsy confides, challenge the very spirit her sister brought to life, a conviction she believed with unwavering trust: "Just do it. Be persistent and you can succeed. Just do it, and never give up. Have the courage to get it done."

In 1980, Nancy founded the Friends of the Columbia Gorge. She recruited an influential board of directors, including Tom McCall and Bob Straub, both former Oregon governors, former Washington State governor, Dan Evans, and Mike Lindberg, Portland City Commissioner. She also engaged Bowen Blair, a young attorney and vice president of Trust for Public Land, to be the Friends' first director. Blair would remain deeply involved in the mission for the rest of Nancy's life.

She also, says Aubrey, employed the power of hard work. "My mother was never idle. She set up an office in our old house in the utility room with a washer and dryer and ironing board. That's where she put her desk and would spend her evenings at the manual typewriter in the laundry room. She was always there until after 10:30 pm almost every night; typing letters, reading. She would take letters to the downtown Central Post Office at night—there was no email—driving down to drop them off because she knew that was the way to have the quickest turnaround. She truly believed that her goal was within the realm of the possible. It was not the work of giants. It's the work of normal people, like us.

"I licked a lot of stamps to put on her envelopes," he adds, laughing.

Aubrey then turns more serious. He begins describing what likely was the most difficult work of all: standing up to abject hostility.

"When you face the kind of real hatred my mom experienced, that's when you are being a truly courageous person," says Aubrey with deep respect.

Bitterness, acrimony, and tension arose from Nancy's efforts to promote awareness of the Columbia River Gorge. To advocate for the creation of a national scenic area, she attended countless hearings and Gorge events hosted by the Friends. These would be met by picketers who didn't want any outsider, least of all coming from Portland, to tell them what they could or couldn't do with their property. Zoning and planning were an anathema, perceived as taking away their rights to do whatever they wanted with their land. They would shout at Nancy, holding signs emblazoned with "Thou Shall Not Covet Thy Neighbor's Land." Cruel songs were written, and records handed out with lyrics such as, "Government, get your hands off our land." Bumper stickers were everywhere to be seen in Skamania County, Washington, with the bold words: "Save the Gorge from Nancy Russell."

Before long, the issue of the scenic area proposal gained national attention. Private-property-rights advocates were brought in from the outside. Many were paid for their disruption. Forces representing the Sagebrush Rebellion were also employed to protest.

The Sagebrush Rebellion, a movement in the 1970s and 1980s, was a highly visible crusade that sought to end federal land control in western states. Supporters believed that the federal government held too much power over western lands, and management of public lands should instead be given to states, local authorities, or become privatized. As much of these areas were in places where sagebrush predominated, the name Sagebrush Rebellion caught on and had cachet, garnering support from individual citizens, politicians, those involved with mineral extraction, gas and oil exploration, and other economic interests. It soon became synonymous for opposition to federal management of public lands. The height of the rebellion, of which Ronald Reagan was a notable champion, lasted only about ten years. It died down in the mid-1980s and lay dormant for close to twenty years. In the twenty-first century, however, the controversy has started up again. For the past decade, a new base of supporters has arisen, attempting to pressure the federal government, sometimes with violent repercussions, to transfer control of publicly owned Western lands to state or private hands.

When Nancy was working to save the gorge with federal legislation in the early 1980s, the Sagebrush Rebellion was at the most heated time of the movement. "People opposing the scenic area act were funded by national organizations and attorneys. They were formidable opponents, with backing by lots of resources," Aubrey relates. "From an emotional point of view, however, it was the local people who were really spouting the personal hatred."

It was Nancy's inner nature, though, to not react to vitriol and scorn with fear. Rather, when things got hard, she responded instead with growing determination.

"My mother was positive. She remained optimistic because she felt she was doing something that wasn't only good in itself, but was joined by so many good people."

Soon, though, Nancy found herself facing a challenge. Very clearly, she understood, its outcome could define the future of the gorge.

In 1981, the year after forming the Friends of the Gorge, Nancy became aware of the gorge's greatest threat to date. On the Washington side of the Columbia, the highly scenic prairie above its defining feature, Cape Horn—a dramatic, massive basalt cliff outcrop—was slated to be developed into fifteen home sites. The lots of Rim View Estates were designated for immediate development and would offer commanding views of the gorge and iconic Multnomah Falls across the river.

Nancy was keenly aware that no zoning existed in Skamania County. She learned that the first and choicest of the properties had already been sold. The rest, now, were up for grabs.

Cape Horn was one of Nancy's most beloved spots. It was bountifully endowed with colorful wildflowers. At different times of the year, tall, purple delphiniums lined the hills, as did quaint, rosy bleeding hearts, native fringe cups, red columbine, and delicate pacific starflowers. The cape held outstanding views of beauty up and down the Columbia River. To Nancy, losing that supreme landscape to rampant development before it could be protected by any national legislation was devastating.

Talking it over with Bruce, who supported her efforts, they made the decision: they must save Cape Horn. To do that, they would mortgage their home. They would immediately take out a bank loan of $400,000, pay the interest themselves, and make the money available to the Trust for Public Land. With this money, the Trust for Public Land would then be able to purchase all of Rim View Estates and save the property and its wild beauty forever.

It was the first time Nancy and Bruce intervened financially with their own funds to protect Gorge property from sprawl. It would not be the last.

Nancy's grassroots labors, which were growing every day, were not escaping the notice of Senator Mark Hatfield. While she advocated at home, he pursued his work for the gorge in Washington, DC. It was not an easy task. If the scenic area act were to be passed, he and Nancy both knew it would have to vault the hurdle of a Reagan administration, well known for its opposition to conservation legislation. Nancy recognized she would have to travel to the nation's capitol herself to make her appeal, and, through a connection, an audience was granted for her with Vice President George H. Bush.

Betsy tells the story with open admiration. "I'm sure my sister's pulse was probably racing considering what she was about to do. The day before she left, she called me and said, 'In less than twenty-four hours I am flying to Washington, DC, and I'm going to lobby to save the gorge! I don't have anything to wear! I don't even have a coat!' Well, that was not surprising," says Betsy knowingly. "My sister hated to shop." She chuckles. "I said to her, 'I have a raincoat, and I use it as a party coat. It's white and quite dressy, and of a fabric that packs well. I'll bring it right over. You can take it.' That's what she took, and practically nothing else."

By every account, Nancy's meeting with Vice President Bush went exceedingly well. Later Betsy saw a picture of her sister, taken when she was sitting in his office. "The picture said it all. Here was my sister, talking to the vice president of the United States! In eight years he would be the president! She was close to the crown then, the royal throne! It was what she was wearing that tickled me. This was the age of chic pant suits and fancy heels, you see. What was my sister wearing? She had on a pair of L. L. Bean loafers, a little dicky blouse with its Peter Pan collar, a plaid

Pendleton skirt, and a cable knit sweater. That was her style. That was Nancy Russell! She could have been wearing that same outfit to Catlin Gabel at age fifteen."

Betsy's amused expression holds a trace of awe. "Here she was, having a conversation with the vice president of the United States of America. You could tell she was perfectly at ease and having the most wonderful time, telling him about this treasure. She liked him. And he liked her! There was an obvious immediate rapport. Vice President Bush probably held the pen for President Reagan when he signed the legislation."

After six years of Nancy's committed leadership and efforts, on November 17, 1986, Congress at last passed the Columbia River Gorge National Scenic Area Act, designating 292,500 acres as federally regulated land. The law would protect the scenic, cultural, recreational, and natural resources of the Columbia River Gorge, while supporting the economy by allowing future growth consistent with that mission.

The act was the only stand-alone conservation legislation ever passed during the Reagan administration.

Nancy Russell and the Friends of the Columbia Gorge had won the battle. The fight to protect the gorge, though, did not end on that day. For as Nancy knew, that occasion was only the beginning. The real goal was to complete the vision: a loop and lacing together of public parks and hiking trails eighty-five miles long on both sides of the Columbia River, from the Sandy River Delta to the west to the Deschutes River to the east.

As Aubrey explains, over the next twenty years, Nancy successfully advocated for purchase into public ownership of forty-thousand acres of gorge lands from willing sellers. She also continued buying properties herself that were threatened with development. In all, she saved thirty-three parcels, totaling hundreds of acres, retaining them for a future time when the US Forest Service or state parks could buy them from her at greatly reduced price. These included key parcels near Mosier, Oregon, sites on the Klickitat River and along Major Creek near White Salmon, Washington, and other locations throughout the gorge.

Aubrey relates that his mother was a one-woman land trust. She nimbly acted, before the Forest Service or state parks could wend through the governmental hoops, to protect properties before they could be snapped up by someone else and developed. With her intervention, she allowed them later to become public spaces.

Today, thirty years after the quest began, a significant portion of Nancy Russell's dream has been realized. There remains more to do, as the land's enduring advocacy group, the Friends of the Columbia Gorge, acknowledges. The flurry of angry dispute, however, has long since passed. Where once flags were flown at half-mast in Skamania County, Washington, when the act was signed into law, it has now been embraced along both sides of the Columbia.

In 2006, the one remaining home on Cape Horn, which later had been purchased by Nancy, was taken down. Native vegetation was planted in its place. Nancy lived to see that glorious day, before passing away in 2008. Since then, Skamania County has built a trailhead at Cape Horn, named in honor of Nancy Russell.

So how did she do it? Betsy ponders. Where did Nancy find the bravery and endurance to not give up?

"For my sister, this was something she just had to do. Nancy was a person who thought of life not as a struggle. She was not one to be swayed from her course or her conviction by intimidation or hardship. To her, there was so much beauty and joy in life! She felt every breath we have is such a privilege. Nancy believed in leaving open land for breathing rather than cluttering it up with man-made buildings. She wanted laws for conservation to be enforced . . . to maintain protection for what exists and then adding more for the future."

Aubrey agrees, then adds a secret he knows about Nancy's character.

"After the Scenic Area Act was passed, and my mother became recognized for her part in it, she was asked to be on a number of national boards. The Smithsonian Institute in Washington, DC, in particular, wanted her. It was a great honor, and probably hard to turn down I would imagine. My mother, though, declined. She knew, in her heart, there was still work to do to protect the gorge. There's a watchdog role that continues today. It is not as glamourous as the Smithsonian, perhaps, but it's the real stuff.

"What did Oregon mean to Mom? It meant everything. While many people might jump to be on that board, or go to Palm Springs for the winter, or have a European vacation every couple years, or any kind of thrill seeking or pleasure seeking, that was not Mom. Oregon was it for her. Oregon was good. It was good enough for her. So why go anywhere else?"

Betsy's tone is one of affection and pleasure as she reflects upon the essence of her sister's life. "While she took things seriously, Nancy loved fun. She loved tennis, and wildflowers and wilderness. She adored her family. And she loved something else: music. My sister was known for always singing off-key, but that never stopped her. There was always music in her home. Music was like sunshine for her. She left her worries at the door. There was never a sadness that would keep her from walking on the sunny side of the street. No matter what, Nancy always walked in the sunshine."

Her eyes take on a glow. "When we went out together, she would grab my arm. Then she'd start to sing, expecting me, of course, to sing with her!"

A smile blooms across Betsy Smith's sweet face, remembering the sister she loved.

"My sister, my soulmate, never said no. She said yes to life."

Afterword

In the ten years I took to write this book and interview scores of people, something became clear. When a place is as rare and beautiful as Oregon, it is worth the effort to try to preserve it.

Then, and now.

Further, the power to change the future lies in people's hearts. It springs from caring for something, and others, beyond yourself.

These people, and many others not recorded, made an impact and improved the quality of all our lives. Sometimes their labors took years, and sometimes decades. To a modern, technology-driven world, such strivings may appear counterintuitive. People may indeed question why anyone would attempt something when there is no assurance of attainment, no prestige, and no monetary gain. Even more, when the odds seem almost impossible.

The inner drive of these Oregonians, however, originated from a stronger and more resilient wellspring for living. It arose from a sense of gratitude, and a personal desire to make their community a better place. Their passion stemmed from a love and appreciation of the natural beauty of Oregon. They wished to see it last beyond their lifetimes. They wished to leave us something good.

Of course, things did not always turn out the way they hoped. Yet this perhaps is the greatest lesson they taught me. I learned that there is something deeper and longer-lasting than merely achieving success. For the sake of humanity, what matters is not just to win.

What matters is to try.

Today we are facing the greatest environmental challenges the world has ever known. We are at a time in our history when leadership and generosity of spirit are needed the most, but seem the scarcest qualities of all.

This, however, is when states have an opportunity to step up to the plate, to make a difference, and to give courage to other states, and individuals, to do the same.

Oregon has been in this role before. Former governor Tom McCall, one of Oregon's most admired and loved governors, inspired generations of people to stand up for the environment with his development of the Bottle Bill, the Beach Bill, and Senate Bill 100. And, just months before he died, he deeply moved Oregonians one final time with words that showed what Oregon meant to him.

It was 1982, and Oregon was in the midst of a recession. Taking advantage of pessimistic opinion, however, opponents of land-use planning, heavily financed by timber interests, real estate developers, and contractors, had gone to work. They had crafted an initiative, called Ballot Measure 6, to repeal Senate Bill 100. Their goal was to do away with zoning "restrictions," putting an end to urban growth boundaries and land-use planning once and for all.

Making matters worse, backers of the measure had successfully convinced the electorate the only way out of state financial woes was to put Senate Bill 100, and all it stood for, on the chopping block. Oregon's visionary law was poised to be extinguished. Of greatest concern to Tom McCall and many others at the time, polls showed the campaign passing by two to one.

Hugely disheartened, before the election McCall attended a small gathering of people opposing Measure 6. Prior to the event, he expressed to the group he had something he wanted to say. Reporters were summoned. During the course of the evening, all awaited with curiosity. When at last he came out to speak, everyone saw something that took them by complete surprise.

The former governor's tall, robust body had become, in only a few months' time, a shadow of its former strength. Greatly weakened, as he rose close to the microphone his face yet held the same resolution for which he was known whenever speaking about his beloved Oregon. Obviously in physical pain, he looked out at the audience rendered to silence.

His earnest words still cut to the heart of all who love this state:

"If you really want to signal to one and all that Oregon is down—that we can't even agree on what we are doing and we are ready to quit—just pass Measure 6 and totally repudiate the Oregon mystique," said Tom Mc-Call. "You all know I have terminal cancer—and I have a lot of it. But what you may not know is that stress induces the spread and induces its activity. Stress may even bring it on. Yet stress is the fuel of the activist. This activist loves Oregon more than he loves life. I know I can't have both very long. The trade-offs are all right with me. But if the legacy we helped give Oregon, and which made it twinkle from afar. . . . If it goes? Well, then, I guess I wouldn't want to live in Oregon anyhow."

Two months later, Tom McCall died. Before he passed away, though, he was able to witness what he had hoped. The day of the election, the people across his state came out, in large numbers, to vote. They did not let him down. Senate Bill 100 remained upheld.

In unified voice, Oregonians overwhelmingly defeated Ballot Measure 6.

This simple yet profound story is the one that inspired me to write this book. It also gives me the hope I need to face our uncertain future. It tells me that if we take up, as Oregonians, the mantle of courage that others before us have carried, together with the genuine desire to safeguard the integrity of this remarkable place for future generations, then what we face is not so bleak after all. Oregon and its people have inspired a nation before.

We can inspire the nation, and the world, once more.

Acknowledgments

I am truly grateful to the many people who shared with me their life stories, feelings, and values that I have tried to faithfully record in *A Generous Nature*. All of those I interviewed provided me with a new and deeper understanding of Oregon, increased my appreciation and affection for this state, and, above all, gave me immense inspiration and hope for the future.

In preparing this book, I spoke to many individuals all across Oregon, and while I was not able to feature all of the interviews in this collection, every conversation offered me insights into what motivates individuals to protect special places. It is my hope that these taped interviews may yet be preserved so that others can listen to the words spoken by the men and women who worked to give us the Oregon we cherish.

I also wish to thank the wonderful supporters of *A Generous Nature*—the book and the vision behind it. I appreciate the Oregon Community Foundation, especially Michael Achterman, Melissa Hansen, Jeff Anderson, and Leann Do, all of whom have been strong advocates throughout. I am especially grateful that the Oregon Community Foundation is graciously donating a copy of the book to every public library in the state. Thanks too to the Oregon Historical Society and executive director Kerry Tymchuk for lending support and backing. I remain indebted to those individuals involved with the Oregon Parks Foundation Fund. They are true champions of Oregon! I am thankful to Kim Stafford for his beautiful poem, "Advice from a Raindrop," which sets the tone for the entire book. Marty Brown and Micki Reaman of Oregon State University Press have been delightful to work with. I appreciate, too, the two outside reviewers who made excellent suggestions to make the book as strong as possible; I eagerly followed their recommendations. I also want to thank Sarah Dugan, my "writing coach," who kept me on task during the years it took to collect all of the interviews and write the stories.

The gorgeous color painting that graces the cover of this book is thanks to the efforts of Mark Humpal, art historian and author, who wrote the beautiful book with stunning plates, *Ray Stanford Strong, West Coast Landscape Artist* (Norman: University of Oklahoma Press, 2017). When I saw the Strong painting of the Columbia River Gorge, I knew it had to be the cover of this book! Mark kindly contacted the owners of the painting, Jeff and Esther Clark, who graciously allowed us permission to use this image.

Of everyone who helped with the manuscript, two people stand above all. Tom Booth, director of Oregon State University Press, guided *A Generous Nature* to the finish line. I cannot thank him enough for his long-standing faith in this book, his hours upon hours of careful review, and his thoughtful and perceptive counsel every step of the journey.

Most of all, I could never have written this book without the support, encouragement, laughter, joy, and tremendous insightfulness of the love of my life—my husband, John Houle. His is the most generous spirit of all.

Additional Resources

The following is a compendium of organizations that found inspiration from the individuals profiled in *A Generous Nature*. These programs, many which had their start from the people in this book, continue to thrive today. This list, though, is only the beginning. There are numerous other groups throughout the state where people are working to protect Oregon's rivers, fish and wildlife, places of beauty, forests, mountains, deserts, seashore, agricultural lands, and wilderness.

With such shared commitment to values, combined with devoted personal involvement and attachment to a piece of ground that touches our souls, we are keeping the Oregon spirit alive. Even more, we are giving a priceless gift to future generations of Oregonians.

Audubon Society of Portland works to promote the understanding, enjoyment, and protection of native birds, other wildlife, and their habitats. It endeavors to protect imperiled species, reduce threats to birds across the Oregon landscape, and to preserve high-priority habitat. (www.audubonportland.org)

Other Audubon chapters:
Lane County Audubon Society (www.laneaudubon.org)
Salem Audubon Society (www.salemaudubon.org)
Cape Arago Audubon Society (www.capearagoaudubon.org)
East Cascades Audubon Society (www.ecaudubon.org)
Corvallis Audubon Society (www.audubon.corvallis.or.us)
Kalmiopsis Audubon Society (www.kalmiopsisaudubon.org)
Klamath Basin Audubon Society (www.klamathaudubon.org)
Rogue Valley Audubon Society (www.roguevalleyaudubon.org)

Siskiyou Audubon Society (www.siskiyouaudubon.org)
Umpqua Valley Audubon Society (www.umpquaaudubon.org)

Coalition of Oregon Land Trusts (COLT) serves and strengthens the land trust community in Oregon. COLT member organizations share a mission to protect, preserve, and steward special lands while working with landowners and various community partners. With twenty-six member organizations, COLT acts to unite conservation nonprofits and become the central voice of the land trust community. (www.oregonlandtrusts.org)

Since 1947, **Defenders of Wildlife** has been dedicated to the protection of all native animals and plants throughout North America in their natural communities. Its work centers on preservation and restoration of imperiled and endangered species and their habitats. It is committed to defending the integrity of the Endangered Species Act. (https://defenders.org)

Eco-School Network is an association of parents working to promote sustainable practices and raising ecological awareness in elementary schools in Oregon. The network equips parents and students to lead school communities toward sustainability through free training and ongoing support. (www.ecoschoolnetwork.org)

Forest Park Conservancy protects and fosters the ecological health of Forest Park. It strives to maintain and enhance the park's extensive trails network and works to inspire community appreciation and future stewardship of this iconic urban forest. (www.forestparkconservancy.org)

The **40-Mile Loop Land Trust** acts to promote and assist in the acquisition of lands along the 40-Mile Loop corridor and utilizes the tools of conservation and recreation easements. (www.40mileloop.org)

Friends of the Columbia Gorge seeks to ensure that the beautiful and wild Columbia Gorge remains a place apart, an unspoiled treasure for generations to come. It strives to vigorously protect the scenic, natural, cultural, and recreational resources of the Columbia Gorge, while promoting responsible stewardship of gorge land, air, and waters. It encourages public ownership of sensitive areas and seeks to educate the public about the unique natural values of the Columbia Gorge. (www.gorgefriends.org)

Friends of Marquam Nature Park is a nonprofit community-based volunteer organization. It works in partnership with Portland Parks and Recreation and individual citizens to conserve, maintain, and enhance the historical, natural, and recreational resources of Marquam Nature Park and to educate the public about the park's unique natural and cultural history. (www.fmnp.org)

Friends of the Metolius strives to protect the Metolius Basin, with its unique qualities of water, forest, wildlife, and spiritual values for future generations. It conducts interpretive, advisory, and educational activities to encourage gentle and respectful human use of the land. (www.metoliusfriends.org)

Friends of Outdoor School is dedicated to increasing student access to Outdoor School programs through advocacy, fundraising, and community engagement. It was instrumental in passing Ballot Measure 99, which secured state funding for a full week of Outdoor School for all fifth- and sixth-grade students in Oregon and continues the fight to fully fund this landmark commitment. (www.friendsofoutdoorschool.org)

Friends of Tryon Creek works in partnership with Oregon State Parks to inspire and nurture relationships with nature in Tryon Creek forest. It promotes community engagement, where people and wildlife thrive in relationship to each other and Tryon Creek State Natural Area. (www.tryonfriends.org)

The High Desert Partnership supports collaboration for restoring healthy forests and wetlands, mitigating wildfire, reshaping youth's experiences, and enhancing the local economy, while fostering these outcomes through conversations and reshaping business as usual. (www.highdesertpartnership.org)

The **Native Plant Society of Oregon** is dedicated to the enjoyment, conservation, and study of Oregon's native plants and habitats. Founded in 1961, it has a statewide network of thirteen chapters with one thousand members. It is involved in local endeavors to protect threatened and endangered species and in re-introduction efforts of native plants. (www.npsoregon.org)

The Nature Conservancy strives to conserve the lands and waters on which all life depends by tackling climate change, protecting land and water, providing food and water sustainability, and building healthy cities. (www.nature.org/en-us)

The Nez Perce Wallowa Homeland serves to secure, develop, and manage Nez Perce lands in a manner that enhances and enriches relationships among the descendants of indigenous people and the contemporary inhabitants of the Wallowa Valley. It works to create a physical place to build these relationships and to support, preserve, and celebrate the customs and culture of the indigenous inhabitants. It assists in assembling the Wallowa Band Nez Perce culture and history in order to provide interpretation, knowledge, and understanding to those who visit the grounds. (www.wallowanezperce.org)

The North Coast Land Conservancy seeks to conserve Oregon's coastal lands, forever. It works to achieve a fully functioning coastal landscape where healthy communities of people, plants, and wildlife all thrive. (www.nclctrust.org)

The **Northwest Earth Institute** endeavors to inspire people to take responsibility for Earth while giving people a framework to talk about their relationship with the planet and to share in discovering new ways to live, work, create, and consume. (www.nwei.org)

Since 1974, **1000 Friends of Oregon** has worked with Oregonians to enhance Oregon's quality of life. It strives to build livable urban and rural communities, to protect family farms and forests, and to conserve natural areas through programs, partnerships, and public involvement. (www.friends.org)

The **Oregon Community Foundation**'s mission is to improve lives for all Oregonians through the service of philanthropy. OCF works with individuals, families, businesses and organizations to create charitable funds that help community causes they care about and to support critical work that nonprofits are doing across Oregon. (www.oregoncf.org)

Oregon Environmental Council advances innovative, collaborative, and equitable solutions to Oregon's environmental challenges. It acts to bring Oregonians together to protect water, air, and land with healthy solutions for today and for future generations. (www.oeconline.org)

As a community-advised fund of the Oregon Community Foundation, the **Oregon Parks Foundation Fund** disburses annual grants supporting the acquisition, preservation, and habitat restoration of Oregon's park land. It works to enhance environmental education and recreational and natural resource improvements to public parks throughout the state of Oregon. (www.oregoncf.org/grants-scholarships/grants/ocf-funds/oregon-parks-foundation)

Oregon Shores is dedicated to preserving the natural communities, ecosystems, and landscapes of the Oregon coast while protecting the public interest in Oregon's beaches established by the Beach Bill. It pursues its mission through education, advocacy, and engaging citizens to keep watch over and defend the Oregon coast. (www.oregonshores.org)

The **Oregon State Parks Foundation** is a statewide, member-supported, nonprofit partner of the Oregon Parks and Recreation Department. It is dedicated to raising funds to enhance and preserve special places and experiences in Oregon's state parks. (www.oregonstateparksfoundation.org)

Oregon Wild works to protect and restore Oregon's wildlands, wildlife, and waters as an enduring legacy for future generations. It endeavors to maintain environmental laws, while building broad community support for protecting wilderness, watersheds, wild and scenic rivers, old-growth forests, and fish and wildlife habitat. It seeks to secure permanent legislative protection for some of Oregon's most precious landscapes. (www.oregonwild.org)

Established in 1981, the **Oregon Wildlife Foundation**'s mission is to empower the lasting conservation of fish and wildlife, and citizen enjoyment of Oregon's natural resources. Working through partnerships with other nonprofits, private industry, membership, and the Oregon Department of Fish and Wildlife (ODFW), it has directed millions of dollars to projects around Oregon aimed at conserving fish, wildlife, and the natural habitat that makes Oregon unique. Members help support conservation, restoration, and education projects around the state. (www.myowf.org)

Through a strong commitment to excellence, accessibility, and good stewardship of Portland parks, the **Portland Parks Foundation** works as a partner to Portland Parks and Recreation to mobilize financial and popular support to ensure a thriving parks system for a healthy Portland. (www.portlandpf.org)

The **Sandy River Watershed Council** works in cooperative partnership with individuals and organizations to improve the health of the Sandy River watershed for fish, wildlife, and people. It coordinates efforts with many private- and public-sector partners to produce the greatest benefits for the watershed, and to protect and restore the natural, cultural, and historic resources of the Sandy River basin. (www.sandyriver.org)

The **Sierra Club, Oregon Chapter** is dedicated to protecting Oregon's wild forests and high deserts, and all public lands. It is active in efforts to influence legislative and administrative decisions that impact Oregon's environment. (www.sierraclub.org/oregon)

SOLVE brings Oregonians together to improve the environment and build a legacy of stewardship. It is dedicated to developing relationships among different groups, individuals, and businesses in pursuit of a common goal: to protect and preserve the places that make up this uniquely beautiful home. SOLVE mobilizes one of Oregon's largest volunteer networks to clean up Oregon's beaches, parks, neighborhoods, and other natural spaces through litter cleanups, invasive plant removal, planting native trees and shrubs, and other environmental projects. (www.solveoregon.org)

The **Southern Oregon Land Conservancy** is dedicated to protecting, preserving, and enhancing the land, wildlife, habitat, and water of the Rogue River region to benefit our human and natural communities. It works to connect people to nature and enrich the quality of life for all who live or visit this spectacular corner of the Pacific Northwest. (www.landconserve.org)

Trust for Public Land is a community service nonprofit that for nearly fifty years has had a mission to create parks and protect natural landscapes for people, with a goal to ensure healthy, livable communities for generations to come. Since 1972, it had been dedicated to connecting communities to the outdoors—and to each other. (https://www.tpl.org)

Wallowa Land Trust is dedicated to preserving the rural nature of the Wallowa Country by working cooperatively with private landowners, Indian tribes, local communities, and governmental entities to conserve land. Wallowa Land Trust uses voluntary, nonregulatory tools for the protection of natural areas, wildlife habitat, open spaces, and working lands. (www.wallowalandtrust.org)

Wallowa Resources works to empower rural northeastern Oregon communities to create strong economies and healthy landscapes through land stewardship, education, and job creation, and by working together with people and community partners. (www.wallowaresources.org)

WaterWatch of Oregon is dedicated to the protection and restoration of Oregon's rivers, streams, and lakes for the benefit of fish, wildlife, and people. It seeks to defend the public interest and to keep regulators accountable. It works to pass balanced water legislation. All WaterWatch efforts have the same high goal: to ensure a legacy of healthy rivers in Oregon (www.waterwatch.org)

Western Rivers Conservancy's motto is "Sometimes to save a river, you have to buy it." WRC acquires lands along outstanding rivers to conserve habitat for fish and wildlife and provide public access for all to enjoy. WRC has created sanctuaries for fish and wildlife and secured recreational access along more than 170 rivers around the West. (www.westernrivers.org)

Wetlands Conservancy is dedicated to promoting community and private partnerships to permanently protect and conserve Oregon's biologically rich and diverse wetlands. It works to conserve, enhance, and restore the physical and ecological values of Oregon's greatest wetlands for current and future generations. (www.wetlandsconservancy.org)

Index

Land is a gift of God.
You are a trustee and the land is the trust.
After we are gone, somebody else must be
the trustee.... I hope they will be there.

———————————

Henry Richmond